INTRO TO PHYSICS
CLASSICAL MECHANICS
COLORING WORKBOOK

ISBN: 9781548482749

COLOR GUIDE

Much of physics involves mathematical formulas. Because it is difficult to portray math in "coloring" form, part of the coloring in this book will be in color-coding the various equations. By assigning a color to each quantity and using those colors throughout the book you'll improve your memory of how the different concepts work. Below are the symbols you'll find in this book and what they stand for. You can use this page to assign a color to each and look back on it for reference as you go through the book.

v - velocity   d - distance

t - time   a - acceleration

g - acceleration of gravitational force

F - force

m - mass   M - mass

G - universal gravitational constant

W - weight   W - work

r - radial distance

E - energy   P - power

K. E. - Kinetic Energy

P. E. - Potential Energy

Physics is the study of matter and energy and how they interact. A lot will be covered in this book! We will start with how **distance**, **velocity** and **time** relate.

Distance is the measurable space between two points.

We will look at a few runners on a race track to practice measuring distance.

FINISH

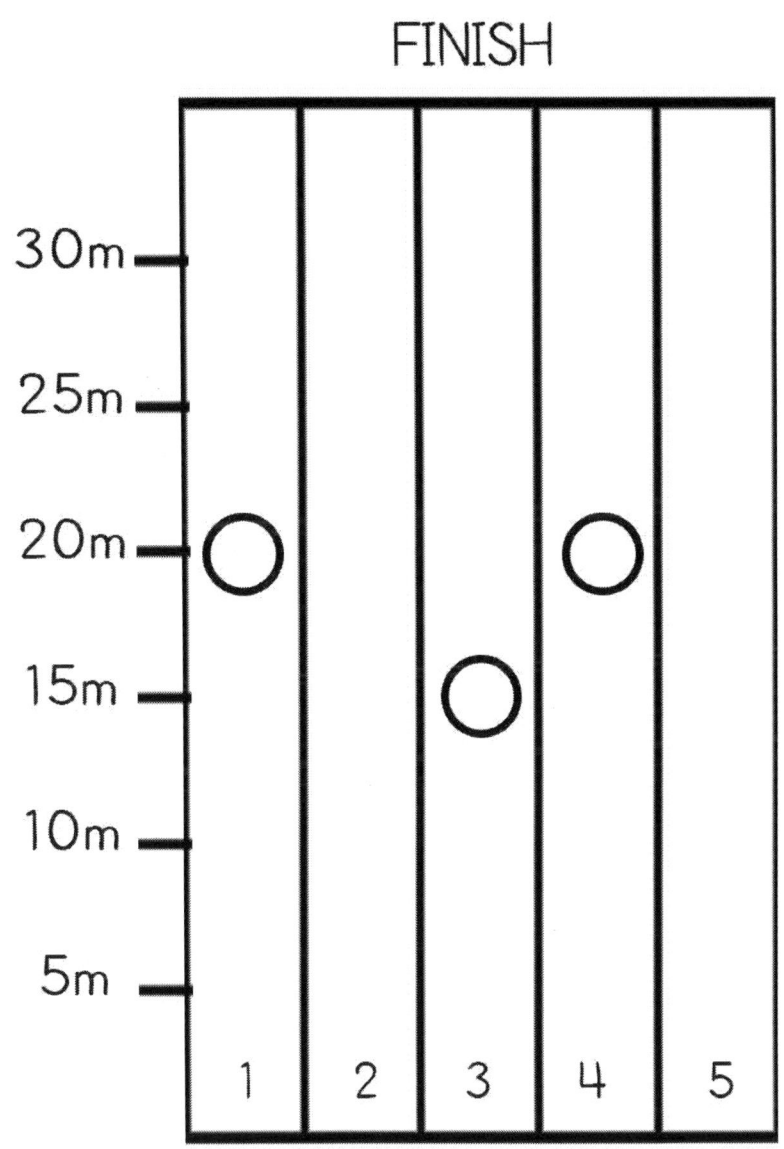

How far is the runner in Lane 1 from the finish line? _____

How much closer to the finish line is the runner in Lane 4 compared to the runner in Lane 3? _____

The runner in Lane 2 is 10m closer to the finsh line than the runner in Lane 3. Draw the runner in Lane 2.

The runner in Lane 5 is 25m from the finish line. Draw the runner in Lane 5.

Time can be defined many different ways. It can be a specific time as seen on a clock, thought of conceptually as in the indefinite passing of time, or, as we'll use it in this book, it can be a specific quantity of time as measured in seconds, minutes, or hours.

It is important to know how to convert time from one measurement to another

1 hour = 60 minutes
1 minute = 60 seconds

To convert hours to minutes, or minutes to seconds, multiply by 60.

1 second = 1/60 minute
1 minute = 1/60 hour

To convert seconds to minutes, or minutes to hours, divide by 60.

How many minutes are in 120 seconds? _____
How many seconds are in 5 minutes? _____
How many minutes are in 1.5 hours? _____
How many hours are in 480 minutes? _____

Velocity is an object's speed and direction. Any time an object changes either speed or direction, it changes velocity. For now we'll focus specifically on speed.

Many of us reference a measurement of velocity on a regular basis when discussing vehicles and traffic!

We might be driving somewhere and see a sign that says

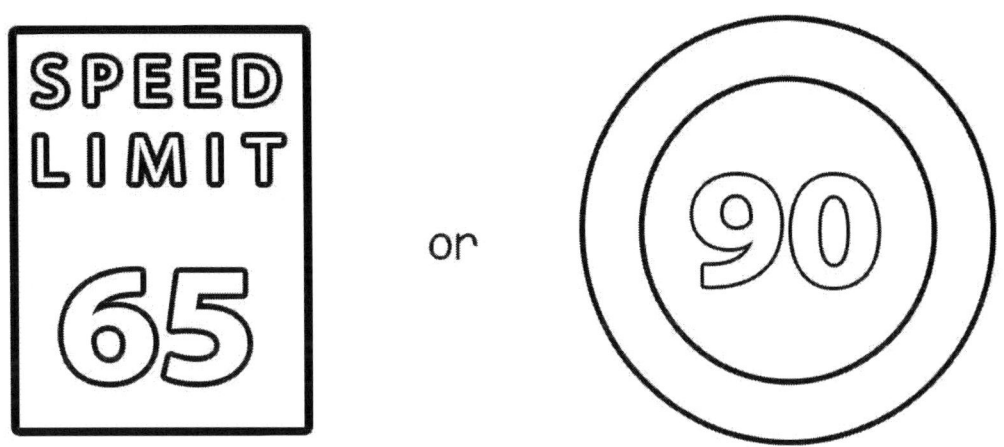

These are two speed limit signs: one from the United States and one from Europe. In the United States, speed limits are given in miles per hour, and in the rest of the world speed limits are given in kilometers per hour. In this book we will be using the metric system for our calculations.

Now, take a moment to think about speed limits again.
Speed limit is measured in **kilometers** per **hour**
In this measurement of **velocity**, we have both a **distance**
and a **time**, and this brings us to our first equation

$$velocity = distance/time$$

$$v = d/t$$

Using some basic algebra, from this equation we can find
two other equations

$$distance = velocity * time$$

$$d = vt$$

$$time = distance/velocity$$

$$t = d/v$$

Back to our racetrack example!

The runner in Lane 2 is 10 meters from the finish line and finishes the race in 3 seconds. What is the runner in Lane 2's velocity?

To solve this, first we look at what we know. We know the **distance** (10m) and we know the **time** (3s), so now we plug these numbers into our equation for finding **velocity**, which is

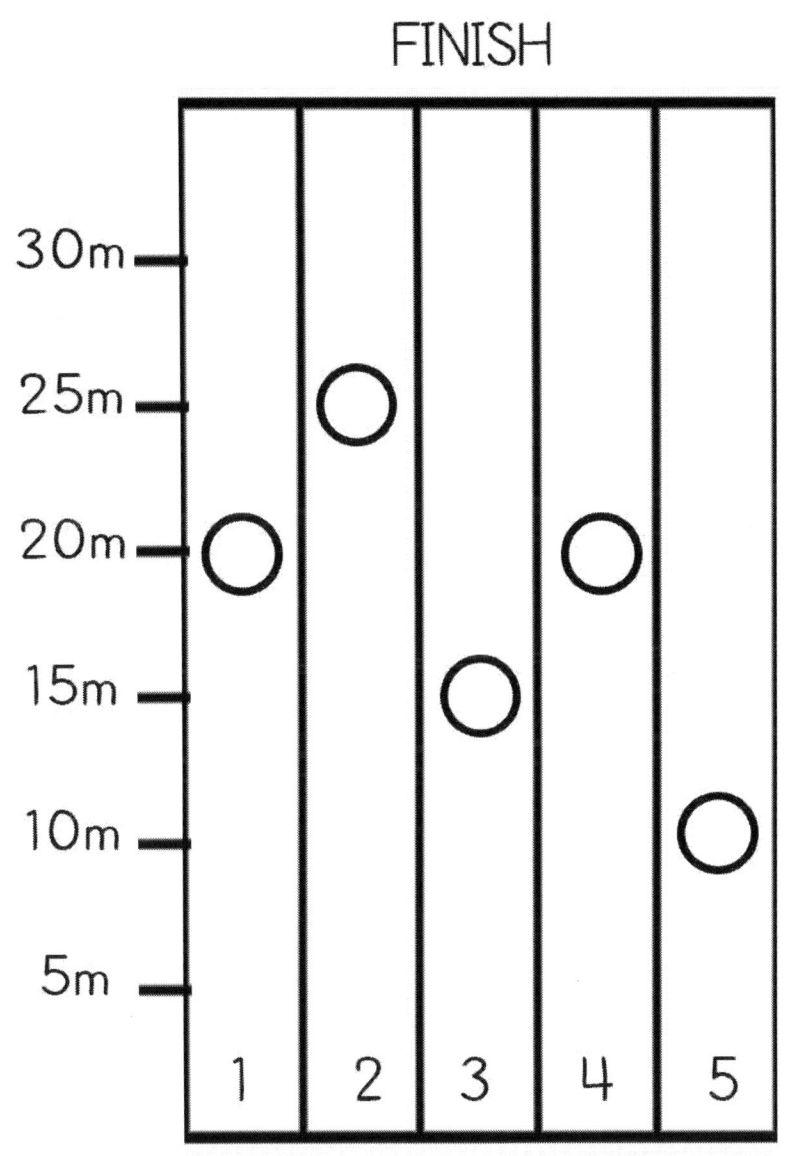

$$v = d/t$$

and we come up with

$$v = \frac{10m}{3s}$$

After dividing 10 by 3 we have an answer of

$$v = 3.33m/s$$

FINISH

30m
25m
20m
15m
10m
5m

1   2   3   4   5

The runner in Lane 5 runs at 2.5 m/s for 6 seconds. How much closer to the finish line is the runner in Lane 5 after these six seconds?

This time we know the **velocity** and the **time**, and we are trying to find **distance**. So we use this equation:

$$d = vt$$

Putting our data into the equation we have

$$d = \frac{2.5m}{s}(6s)$$

And after multiplying, our answer is

$$d = 15m$$

The seconds cancel each other out, leaving us an answer in meters. The runner in Lane 5 will be 15 meters closer to the finish line after running at a pace of 2.5 m/s for 6 seconds.

We've done an equation finding **velocity** and one finding **distance**, so now we'll find **time**.

The runner in Lane 1 is running at a pace of 3m/s. How long will it take the runner in Lane 1 to finish the race?

To solve this, we'll use the equation

$$t = d/v$$

Since velocity is a fraction, meters divided by seconds, to divide by velocity we multiply by its inverse.

$$t = \frac{15m}{1}\left(\frac{s}{3m}\right)$$

Meters cancel out and after dividing we are left with an answer of:

$$t = 5s$$

FINISH

30m

25m

20m

15m

10m

5m

| 1 | 2 | 3 | 4 | 5 |

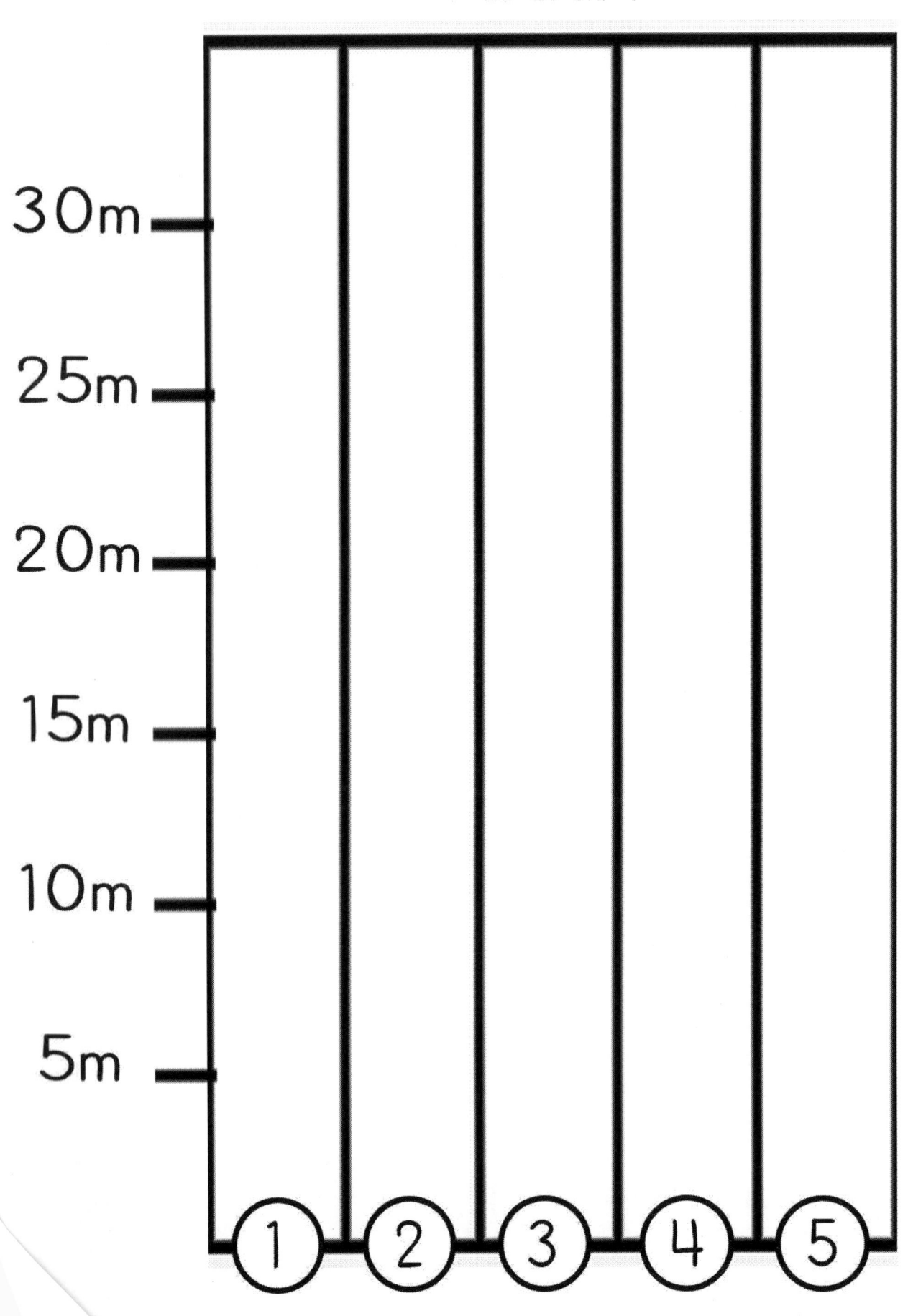

FINISH

30m

25m

20m

15m

10m

5m

① ② ③ ④ ⑤

8

Now for some practice questions!

All runners are at the starting line.
Runner 1 starts running at a pace of 5 m/s
Runner 2 starts running at a pace of 4.5 m/s
Runner 3 starts running at a pace of 6 m/s
Runner 4 starts running at a pace of 3.5 m/s
Runner 5 starts running at a pace of 6.5 m/s

Using **d=vt**, how far are each of the runners after 2 seconds? Use the space below to solve for **d**, then draw the runners in their new locations on the race track.

Runner 1   Runner 4

Runner 2   Runner 5

Runner 3

For the next 3 seconds of the race, each of the racers change velocity. Use **v=d/t** to solve for each racer's new velocity, and tell whether they are running faster or slower than before. Draw their new locations on the race track. Runner 1 makes it to the 20m line.

Runner 2 makes it to the 25m line.

Runner 3 makes it to the 25m line.

Runner 4 makes it to the 20m line.

Runner 5 makes it to the 30m line.

Now all of the racers have finished the race! Use **t=d/v** to find how many seconds it took each of them to complete the final portion of the race. Then tell which order the racers were in when they crossed the finish line.

Runner 1 finishes at a velocity of 3.23 m/s

Runner 2 finishes at a velocity of 4.65 m/s

Runner 3 finishes at a velocity of 5.13 m/s

Runner 4 finishes at a velocity of 2.95 m/s

Runner 5 finishes at a velocity of 4.16 m/s

1st place _____   4th place _____
2nd place _____   5th place _____
3rd place _____

Our next topic is **acceleration**. Acceleration is a change in velocity over time. As an equation this can be written a few ways. One way to represent "change in" is a triangle.

$$\text{acceleration} = \frac{\text{change in velocity}}{\text{time}}$$

$$a = \frac{\text{change in } V}{t} \qquad a = \frac{\Delta V}{t}$$

$$a = \frac{\text{final } V - \text{initial } V}{t}$$

Changing the equation to find time or the change in velocity, we have:

$$\text{change in } V = at$$

$$t = \frac{\text{change in } V}{a}$$

For our next examples we'll look at a car driving on a straight road.

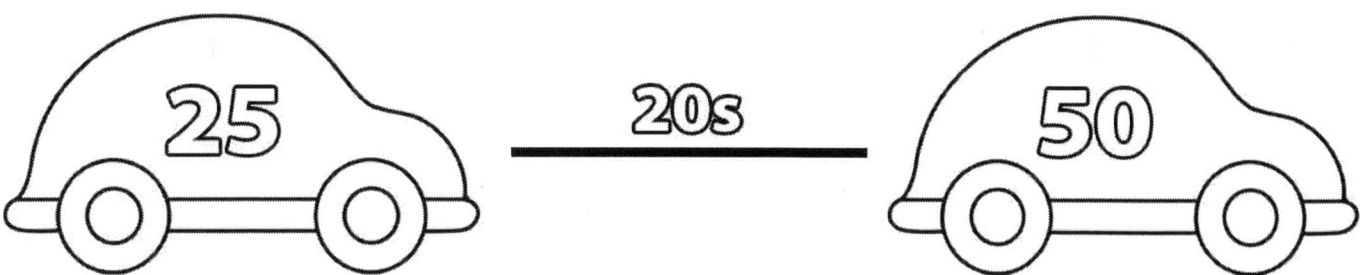

The car starts at a velocity of 25 km/hr and 20 seconds later is driving at a velocity of 50 km/hr. What is the rate of acceleration?

First we find the **change in velocity**

$$\text{final } V - \text{initial } V = \text{change in } V$$

$$50\text{km/hr} - 25\text{km/hr} = 25\text{km/hr}$$

Then we put the numbers into our equation.

$$a = \frac{25\text{km/hr}}{20\text{s}}$$

Now, you might notice our units are different. The equation shows both hours and seconds. We need to make both units of measurement the same to complete the problem.

13

To change the unit from hours to seconds, we add

$$\left(\frac{1\ hr}{3600s} \right)$$

to the equation.

To change the unit from kilometers to meters, we add

$$\left(\frac{1000m}{1km} \right)$$

to the equation.

Giving us the final equation of:

$$a = \frac{25\ km/hr}{20s} \left(\frac{1\ hr}{3600s} \right) \left(\frac{1000m}{1km} \right) = 0.347\ m/s^2$$

Notice we have an answer with a unit of meters per second squared. If our velocity increases by 25km/hr over the course of 20 seconds, our rate of acceleration is 0.347 meters per second per second!

This time we know the acceleration but not the time.

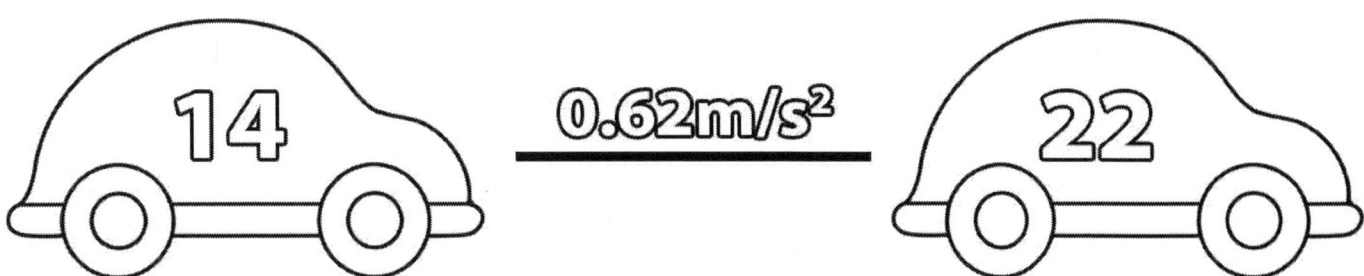

The car starts at a velocity of 14 km/hr and accelerates at a rate of 0.62m/s^2 up to a velocity of 22 km/hr. How much time passed?

First we find the **change in velocity**

$$\text{final } V - \text{initial } V = \text{change in } V$$

$$22 \text{km/hr} - 14 \text{km/hr} = 8 \text{km/hr}$$

Then we put the numbers into our equation.

$$t = \frac{8 \text{km/hr}}{0.62 \text{m/s}^2}$$

Once again, we need to make both units of measurement the same to complete the problem.

To cancel out the units in kilometers, meters, and hours, and to go from seconds squared to seconds, once again we add:

$$\left(\frac{1\ hr}{3600s}\right)\left(\frac{1000m}{1km}\right)$$

Giving us a final equation of:

$$t = \frac{8km/hr}{0.62m/s^2}\left(\frac{1\ hr}{3600s}\right)\left(\frac{1000m}{1km}\right) = 3.58s$$

Now we see that if we accelerated at a rate of $0.62m/s^2$ it took 3.58s to increase our velocity by 8km/hr.

Here is a more expanded version of the same equation so you can see how the units cancel out.

$$t = \left(\frac{8km}{hr}\right)\left(\frac{s^2}{0.62m}\right)\left(\frac{1hr}{3600s}\right)\left(\frac{1000m}{1km}\right)$$

Now we will find the change in velocity.

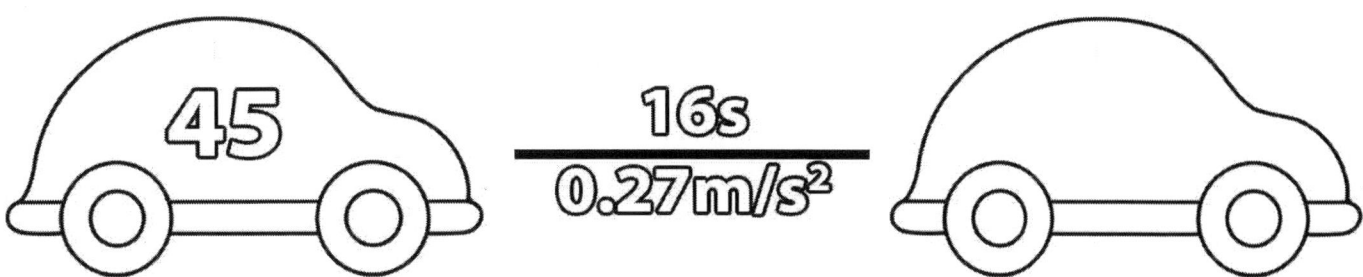

The car starts at a velocity of 45 km/hr and accelerates at a rate of 0.27m/s² for 16 seconds. What is the final velocity?

Right away we can plug our numbers into the equation:

$$\text{change in } V = 0.27m/s^2(16s) = 4.32m/s$$

Now we change the units from meters per second to kilometers per hour and calculate the final velocity

$$\frac{4.32m}{s}\left(\frac{3600s}{1hr}\right)\left(\frac{1km}{1000m}\right) = \frac{15.55km}{hr}$$

$$45km/hr + 15.55km/hr = 60.55km/hr$$

Time for some practice questions! Look at the information you have and solve for the unknown.

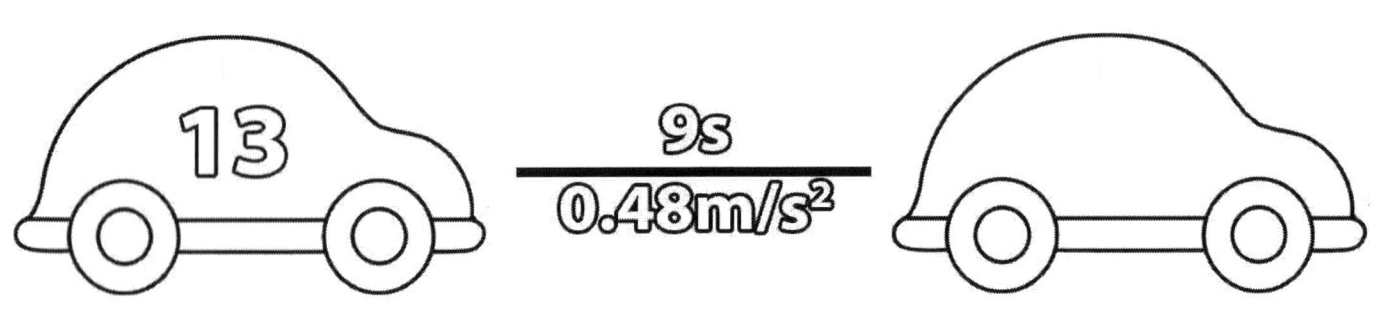

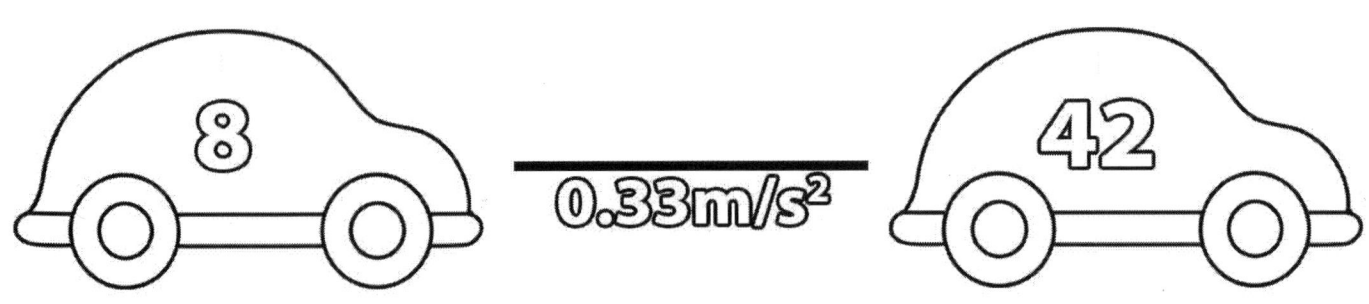

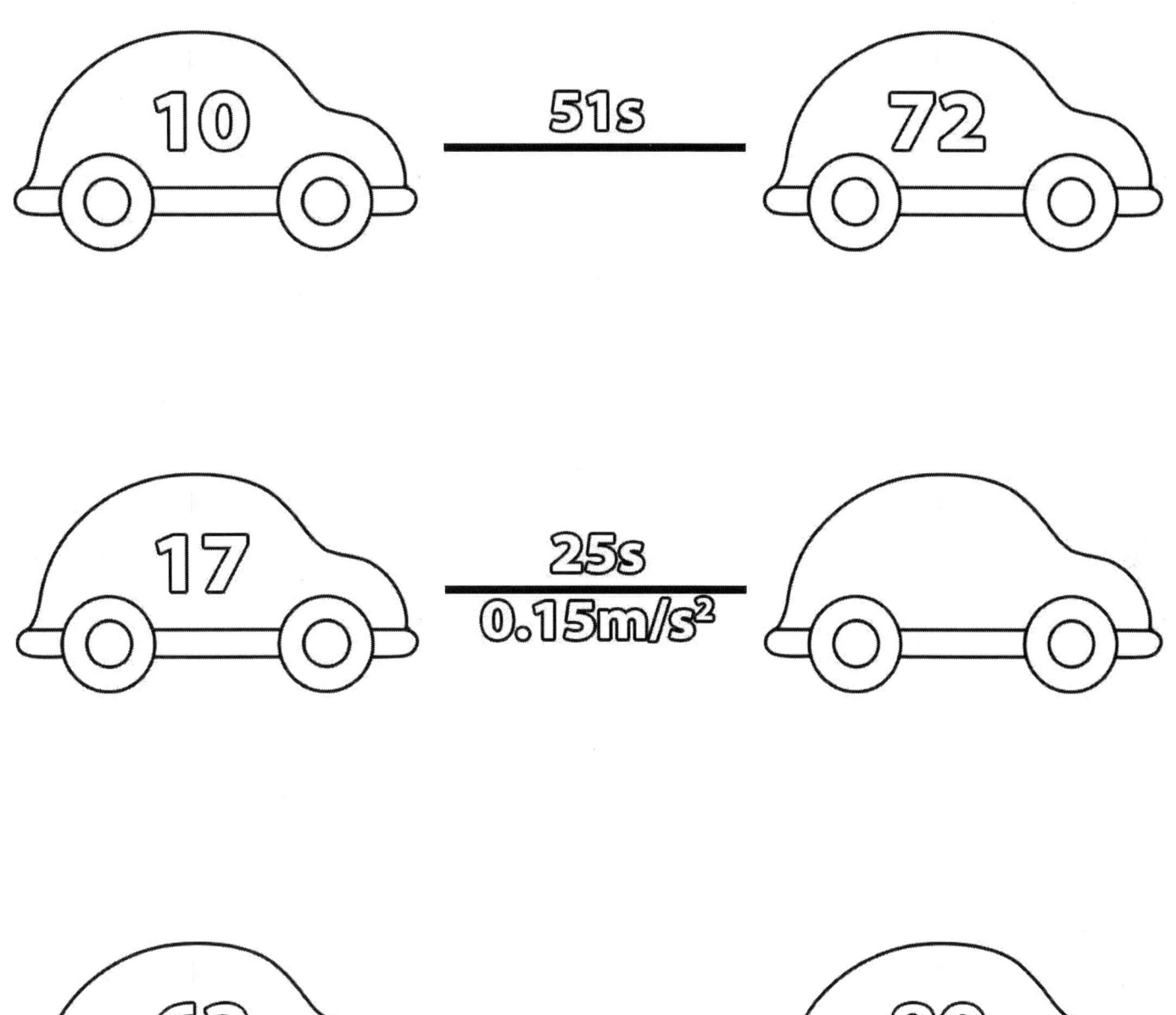

We've looked at how **distance**, **time**, and **velocity** relate, and how **acceleration**, **time**, and **velocity** relate. Now we'll look at how to find **distance** using **acceleration** and **time**!

We know that $d = vt$ and $\triangle v = at$

The velocity we're looking at in **d=vt** is **average velocity**. In $\triangle$**v=at** we have the **change in velocity**. To use acceleration and time to find distance, we have to find their relation to the average velocity.

Change in velocity is the difference between the final velocity and the starting velocity

$$\triangle V = V_{final} - V_{start}$$

Average velocity is the starting velocity plus the change in velocity, divided in half

$$V_{average} = V_{start} + \frac{\triangle V}{2}$$

So:

If we know acceleration and time, we know the change in velocity.

$$\triangle v = at$$

If we know the change in velocity and the starting velocity, we can figure out the average velocity.

$$V_{average} = V_{start} + \frac{\triangle V}{2}$$

If we know the average velocity and time, then we know distance.

$$d = V_{average}\, t$$

Therefore, if we know acceleration, time, and starting velocity, we can find distance.

$$d = (V_{start} + \frac{1}{2}at)t$$

or

$$d = (V_{start})t + \frac{1}{2}at^2$$

Now, some basic algebra gives us the equations to determine time and acceleration.

To solve for acceleration...:

$$a = \frac{2(d - v_{start}t)}{t^2}$$

Solving for time is slightly more complicated. First we move d over to the other side of the equation:

$$\frac{1}{2}at^2 + v_{start}t - d = 0$$

and then you'll see that this is a quadratic formula. The quadratic equation is **$ax^2 + bx + c = 0$**. To solve for t we use the following:

$$t = \frac{-v_{start} + \sqrt{v_{start}^2 + 2ad}}{a}$$

If starting velocity is zero, our equations become:

$$d = \frac{1}{2}at^2$$

$$a = \frac{2d}{t^2}$$

$$t = \sqrt{\frac{2d}{a}}$$

Now for a few examples of this formula group in practice. If an object starts from rest and accelerates for 6 seconds at a rate of $5.2 m/s^2$, what distance has been covered? We'll use the following formula for distance:

$$d = \frac{1}{2}at^2$$

and after plugging in our numbers we have:

$$d = \frac{1}{2}(5.2m/s^2)(6s)^2 = 93.6m$$

We'll use those same numbers to check our answer with the other equations.

We know that:

$a = 5.2 \text{m/s}^2$ $t = 6\text{s}$ $d = 93.6\text{m}$ starting velocity = 0

Let's solve for **a** and see if the math works out!

$$a = \frac{2d}{t^2}$$

$$a = \frac{2(93.6\text{m})}{6^2} = 5.2 \text{m/s}^2$$

It checks out! Now let's try again but solving for t.

$$t = \sqrt{\frac{2d}{a}}$$

$$t = \sqrt{\frac{2(93.6\text{m})}{5.2\text{m/s}^2}} = 6\text{s}$$

Now try a few problems yourself using these equations.
If we know an object started at zero velocity, accelerated for 5 seconds and covered a distance of 12 meters, what was the rate of acceleration?

If an object starts at zero velocity and accelerates at a rate of 1.34m/s^2, how long does it take to travel 16 meters?

If an object starts at zero velocity and accelerates at a rate of 4.6m/s^2 for 11.8 seconds, how far did it go?

So far we've used examples with a starting velocity of zero. The equations are slightly more complex when we have a different starting velocity, and now we'll look at that.

To review, the equation to solve for distance when we know acceleration, time, and starting velocity, is:

$$d = (V_{start})t + \frac{1}{2}at^2$$

We'll use the following numbers:

starting velocity = 1.3m/s   t = 14.1s   a = 0.92m/s²

$$(1.3m/s)14.1s + \frac{1}{2}0.92m/s^2(14.1s)^2$$
$$d = 109.78m$$

Let's test this answer with the other equations.

Again, the equation to find acceleration with time, distance, and starting velocity, is:

$$a = \frac{2(d-V_{start}t)}{t^2}$$

Plugging in the numbers, including our answer for **d** from the previous question...

$$a = \frac{2(109.78m - 1.3m/s(14.1s)}{(14.1s)^2}$$

$$a = \frac{2(109.78m - 18.33m)}{198.81s^2}$$

$$a = \frac{2(91.45m)}{198.81s^2}$$

$$a = \frac{182.9m}{198.81s^2}$$

$$a = 0.92m/s^2$$

Success!

And now to solve for time.

$$t = \frac{-V_{start} + \sqrt{V_{start}^2 + 2ad}}{a}$$

$$\frac{-1.3m/s + \sqrt{(1.3m/s)^2 + 2(0.92m/s^2)109.78m}}{0.92m/s^2}$$

$$\frac{-1.3m/s + \sqrt{1.69m^2/s^2 + 201.99m^2/s^2}}{0.92m/s^2}$$

$$\frac{-1.3m/s + \sqrt{203.68m^2/s^2}}{0.92m/s^2}$$

$$\frac{-1.3m/s + 14.27m/s}{0.92m/s^2}$$

$$\frac{12.97m/s}{0.92m/s^2}$$

$$t = 14.1s$$

Now you'll try a few on your own.

An object moving 10.8m/s begins accelerating at a rate of 0.16m/s^2 and travels for 24 seconds. What distance is covered in this time?

Using your answer for distance, pretend you don't know acceleration. Go through the steps to find acceleration and check your work.

Now do the same to find time.

Review!

These are the formulas we've learned so far

$$d = vt \qquad v = d/t \qquad t = d/v$$

$$\Delta v = at \qquad a = \Delta v/t$$
$$t = \Delta v/a$$

$$d = \tfrac{1}{2}at^2$$

$$a = \frac{2d}{t^2} \qquad t = \sqrt{\frac{2d}{a}}$$

$$d = (v_{start})t + \tfrac{1}{2}at^2$$

$$a = \frac{2(d - v_{start}t)}{t^2}$$

$$t = \frac{-v_{start} + \sqrt{v_{start}^2 + 2ad}}{a}$$

A rain cloud is 600 meters from your house and traveling toward it at a velocity of 0.68m/s. If velocity remains constant, how long will it take the rain cloud to reach your house?

Circle the types of information you have and underline the type of information you need. Then write the formula you should use and solve the equation:

distance | time | starting velocity | final velocity | average velocity | $\triangle$velocity | acceleration

After traveling 50 meters toward your house, the rain cloud begins accelerating at a rate of $0.07m/s^2$ Now how long will it take to reach your house?

distance | time | starting velocity | final velocity | average velocity | $\triangle$velocity | acceleration

What is the rain cloud's velocity after the first 28 seconds of acceleration?

distance | time | starting velocity | final velocity | average velocity | $\triangle$ velocity | acceleration

How far has it traveled now? How much further does it need to go to reach your house?

distance | time | starting velocity | final velocity | average velocity | $\triangle$ velocity | acceleration

The rain cloud's acceleration increases. It will reach your house in 38 seconds. What is the new rate of acceleration?

distance I time I starting velocity I final velocity I average velocity I $\triangle$ velocity I acceleration

What is the average velocity during this time?

distance I time I starting velocity I final velocity I average velocity I $\triangle$ velocity I acceleration

You are 72 meters from your house. The rain will reach your house in 38 seconds. How fast do you have to run to get home before the rain?

distance I time I starting velocity I final velocity I average velocity I $\triangle$ velocity I acceleration

The cat sees a bird in the tree and runs toward it. The cat starts from rest and accelerates at a rate of 0.5m/s^2, arriving at the tree in 3.8 seconds. How far did the cat run?

distance | time | starting velocity | final velocity | average velocity | $\triangle$velocity | acceleration

The bird, seeing the cat, takes off from the tree and flies to the swingset. From the tree to the swingset is 15m and the bird arrives in 4.3 seconds. What was the bird's rate of acceleration?

distance | time | starting velocity | final velocity | average velocity | $\triangle$velocity | acceleration

The dog, seeing the cat at the tree, moves toward it at a velocity of 1.8m/s, arriving in 2.6 seconds. What was the distance between the dog and the tree?

distance | time | starting velocity | final velocity | average velocity | $\triangle$velocity | acceleration

Mark was riding his bike to a friend's house. When he got to the fire hydrant he was moving 2 meters per second. 16 seconds later when he got to the library he was riding 4 meters per second. What was his acceleration during this time?

distance | time | starting velocity | final velocity | average velocity | $\triangle$ velocity | acceleration

If his acceleration above was constant for the entire bike ride, and it's 100 meters from his house to his friend's house, how much time did it take him to make it there?

distance | time | starting velocity | final velocity | average velocity | $\triangle$ velocity | acceleration

He stopped at the library on the way home. His starting velocity when he left his friend's house was zero, and his velocity when he got to the library was 3m/s. His acceleration was 0.5m/s². How much time did it take him to get to the library?

distance | time | starting velocity | final velocity | average velocity | $\triangle$ velocity | acceleration

37

On page 3 we defined **velocity** as an object's **speed** and **direction**. So far we've only been using examples where direction is constant. Now we'll look at the role direction plays in velocity and acceleration.

To illustrate how direction matters, we must view velocity and acceleration as vector qualities.

A vector is a quantity with more than one element, with both direction and magnitude, represented by an arrow.

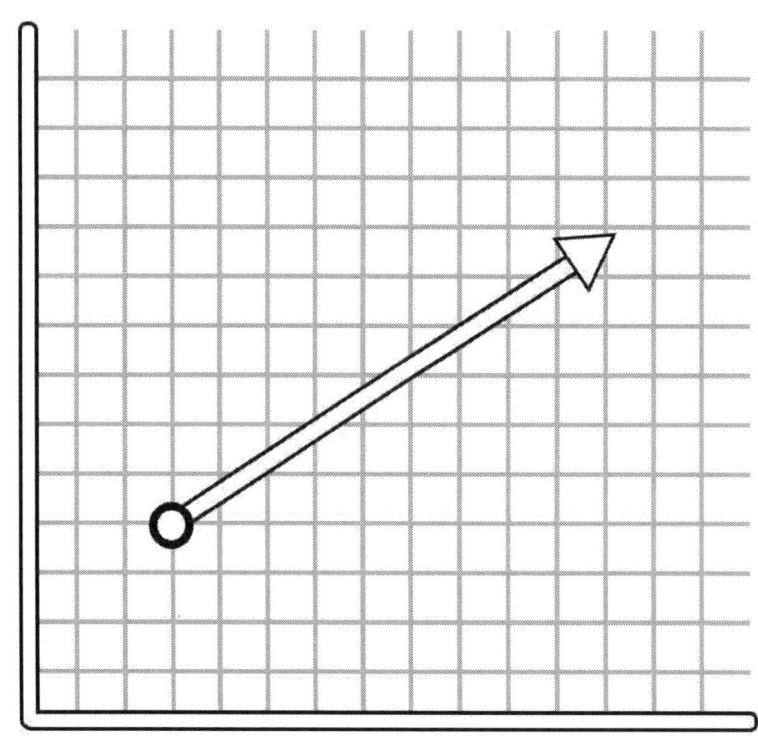

With velocity, the direction the object is going is its direction, and its speed is its magnitude.

If the direction of movement changes, the direction of the vector changes, and therefore the velocity has changed, even if speed remains the same. It is the same for acceleration. Because of this, if an object begins moving in reverse, it is said to have negative velocity, and if an object begins slowing down, it is said to have negative acceleration.

Consider the vector at right and imagine that it represent's a car's velocity as it drives forward out of a garage at a constant speed.

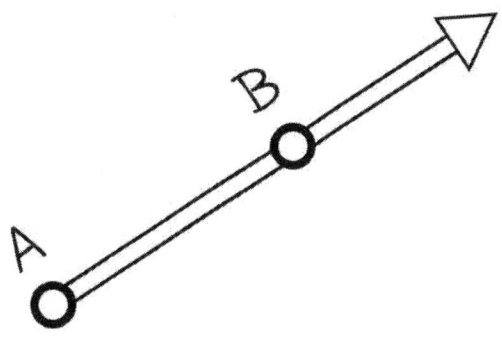

When the car gets to point B, the driver realizes he forgot something in the house, so he backs the car into the garage again, back to point A where he started. Because the vector is pointing in a specific direction and the car is now moving opposite that direction, even though the car has speed and is moving, it has negative velocity.

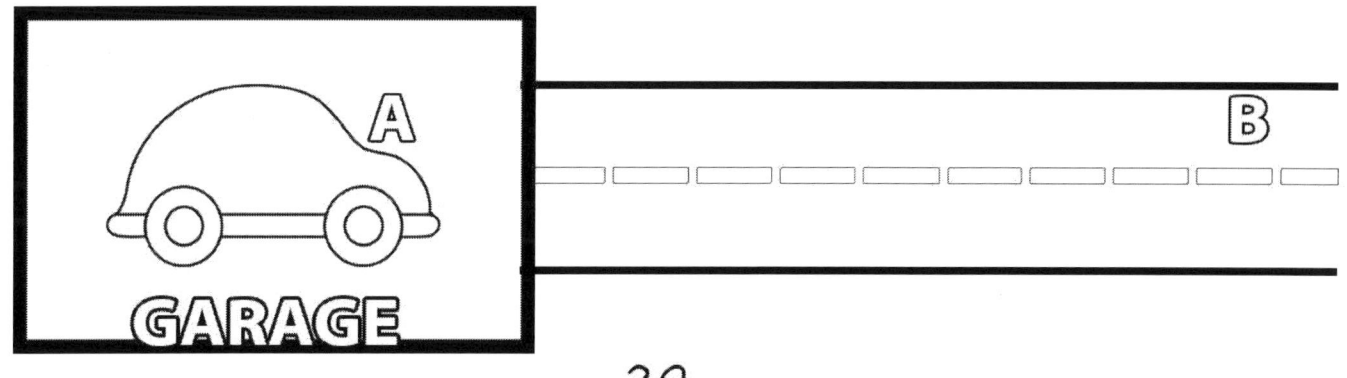

Consider also a car driving at a constant speed around a circular track. While the speed stays constant, the velocity is continually changing because the direction is changing.

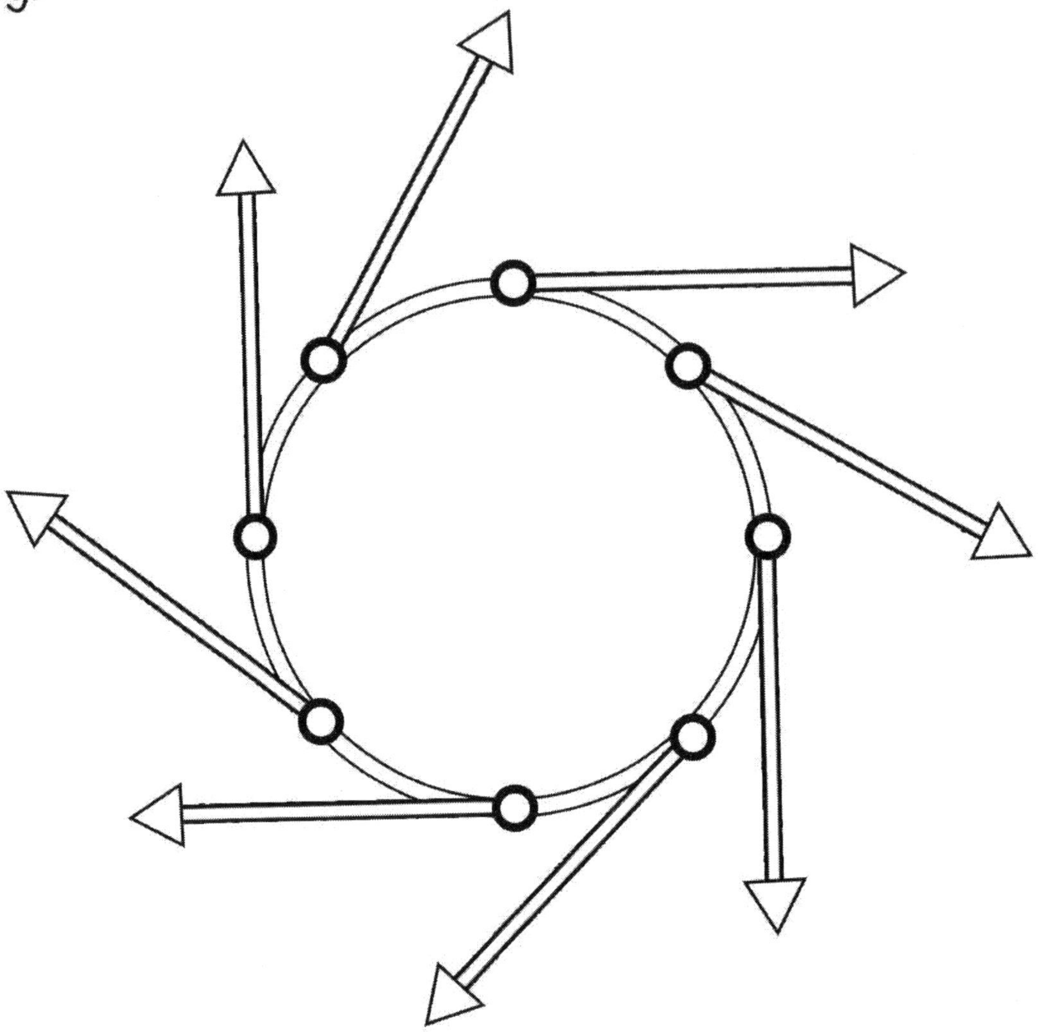

This diagram shows vector representations at 8 points around the circle. Because the car is not moving in a straight line, following any vector point along the circle's circumference, it has a constantly changing velocity regardless of speed. Color each vector a different color to represent the different velocities.

Acceleration is also a vector quality. When an object has constant speed it has zero acceleration, when it is gaining speed, it has positive acceleration, and when slowing down it has negative acceleration. Looking at it on the points of a vector, positive acceleration is like moving from point A toward point B, and negative acceleration is moving back from point B toward point A, only instead of the direction of movement, you're looking at the direction of velocity.

In positive acceleration, the object's velocity is moving further from zero, toward point B.
In negative acceleration, the object's velocity is moving closer to zero, toward point A.

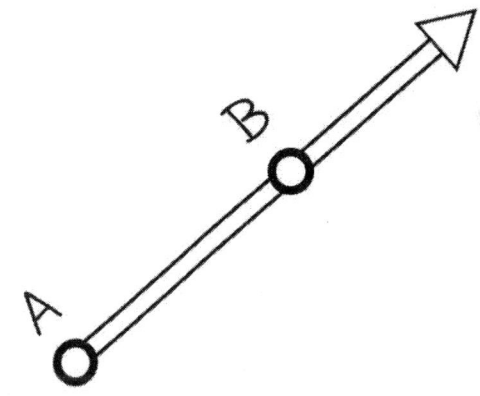

On the next pages we'll look at the acceleration and velocity of a ball as it moves on a flat plane, then down and up a curve.

A ball rolling at a constant velocity then rolls down and up the edges of a curve. What happens to the ball's acceleration and velocity?

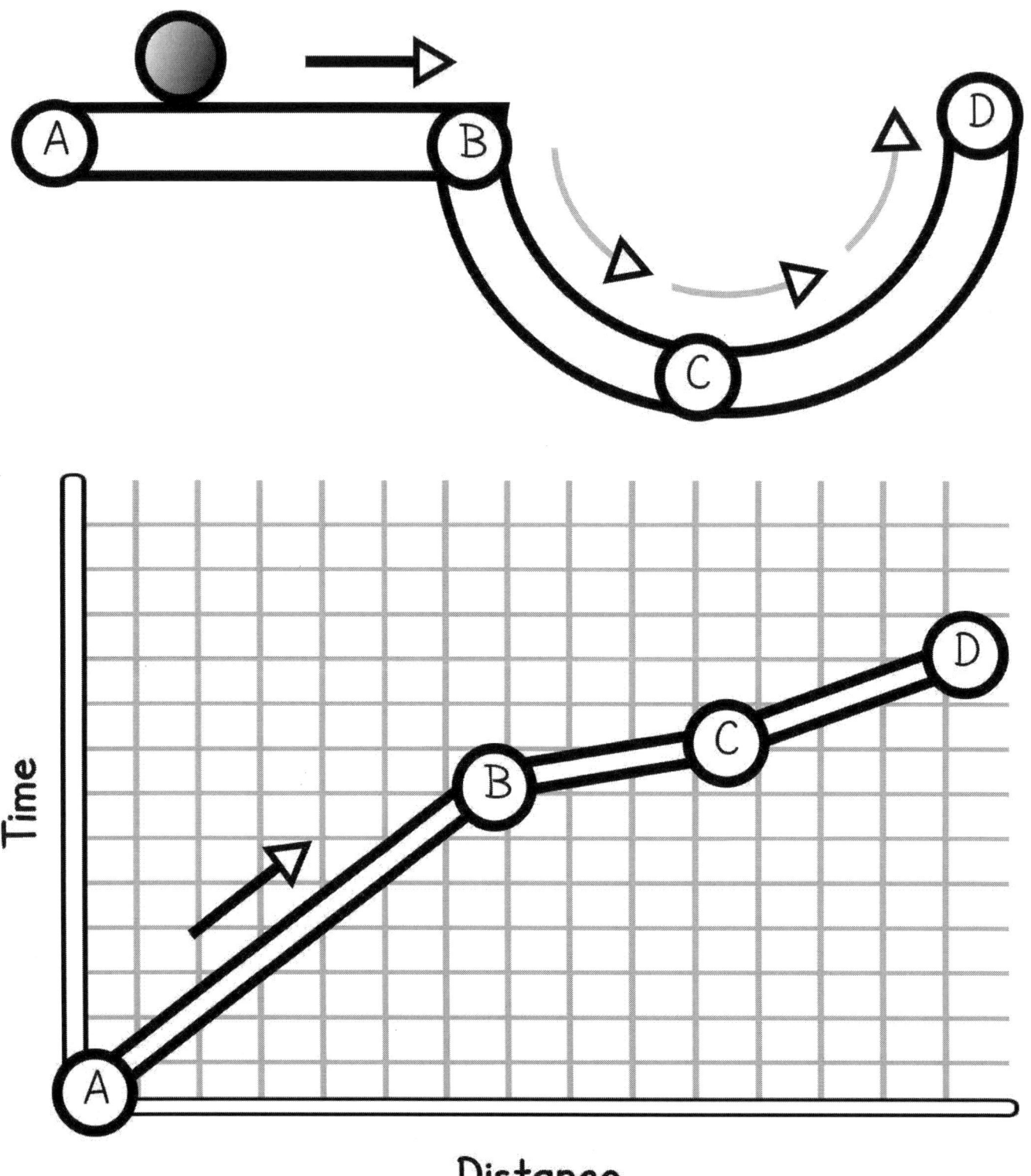

Time

Distance

Looking at this graph we have Distance on the X axis and Time on the Y axis. Because Time only moves in one direction and does not reverse, the point on the Y axis will continually climb upward. Time is not a vector

Distance can be viewed one of two ways. It can be seen as the literal amount of space the ball touches as it rolls, which like time will continually increase, until the ball stops moving, or it can be seen as the displacement between starting point and ending point. Displacement is why velocity is a vector

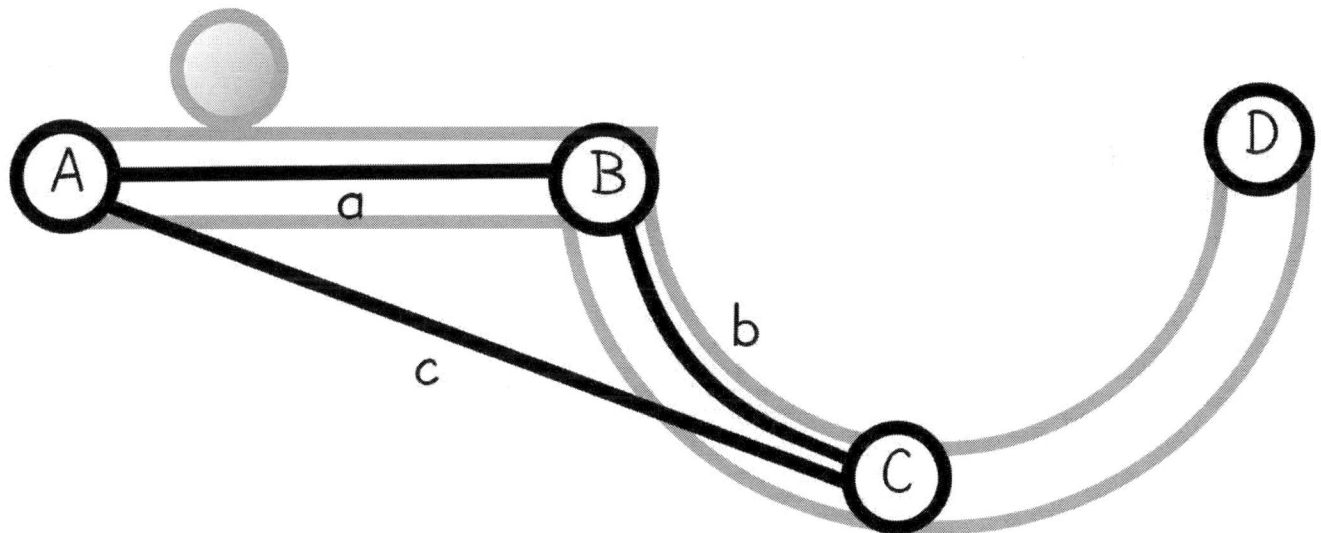

If the length of a + b is distance traveled, the length of c is displacement. This is why if the ball were to roll backwards it would have negative velocity: the displacement would be shrinking.

When the ball moves from A to B to C to D, the velocity changes but remains positive. But if the ball were to come to a stop at point D and roll back down to C, we'd then have negative velocity as displacement moves closer to zero.

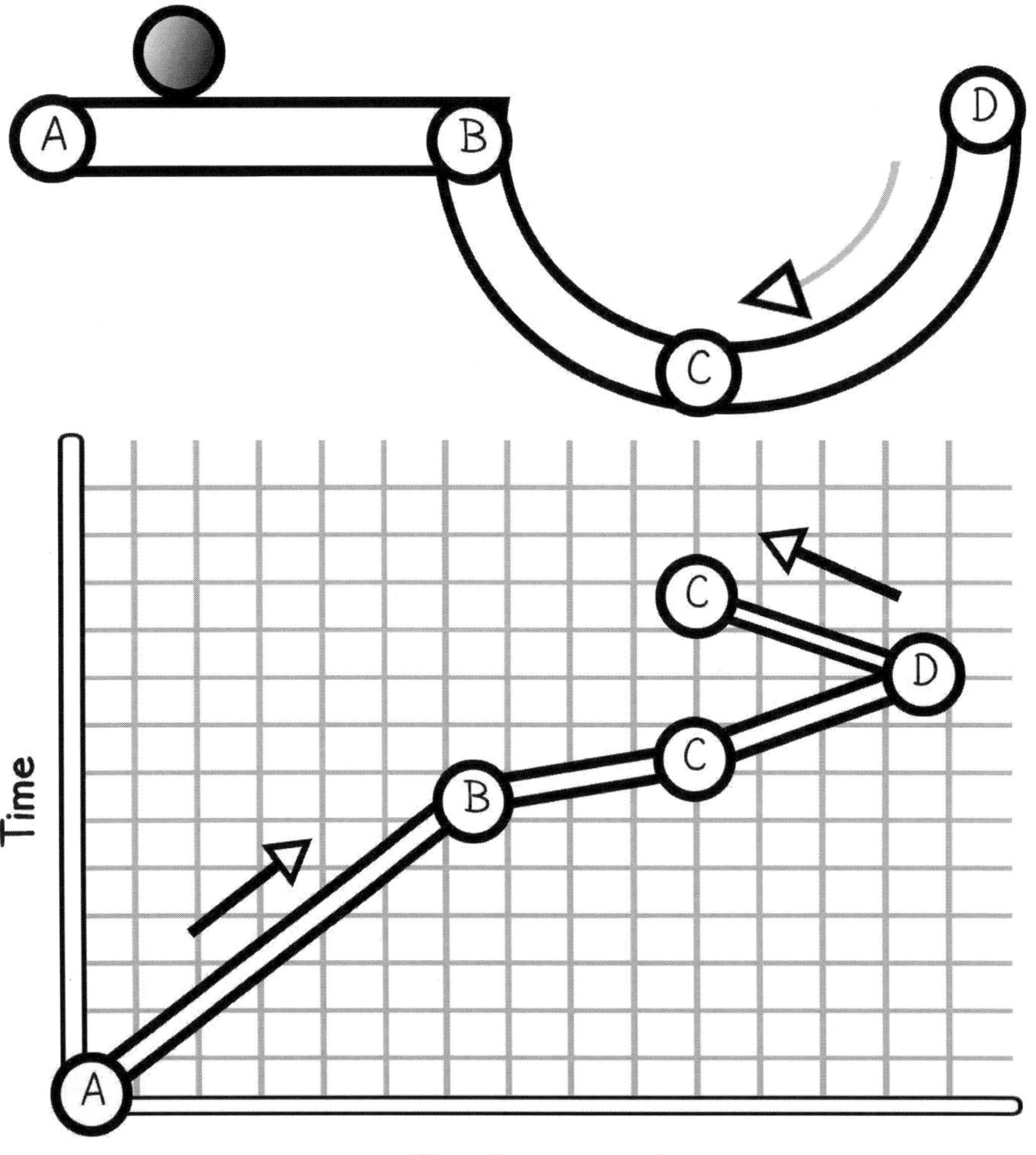

Time

Displacement

44

As you can see, time continues increasing, but now the distance/displacement is decreasing. So from point A to point D we have positive velocity, and then from point D back to point C we have negative velocity. Shown below, when the ball rolls back to point C, displacement changes from being the length of line d to the length of line c.

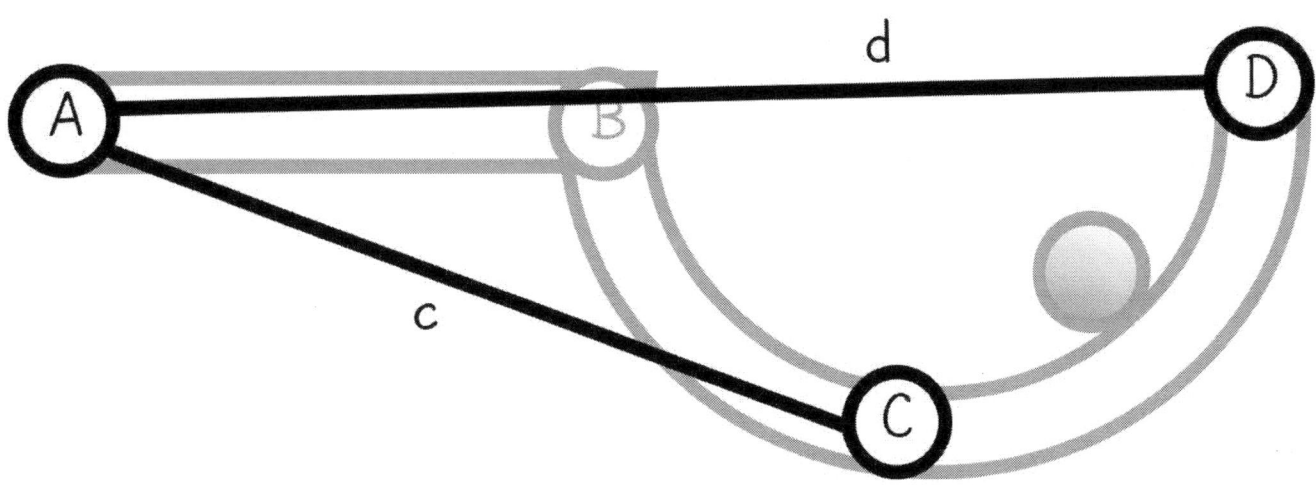

If the ball were to continue rolling backwards until it arrived at point B again, approximately where might that point be on the graph? Draw this. Does the ball have positive or negative velocity? _____

Now we'll look at the same ball but consider its acceleration instead of velocity.

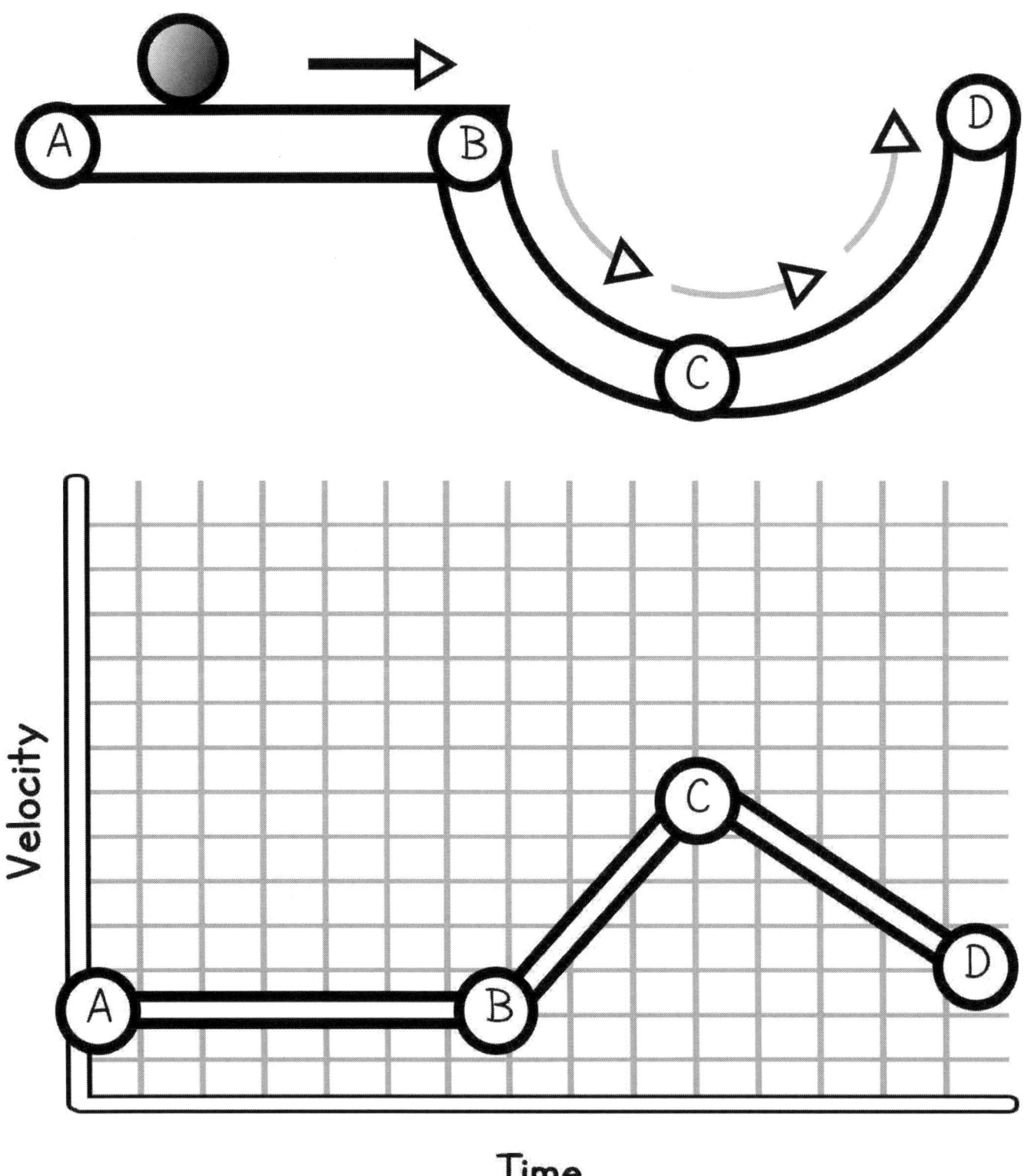

Velocity

Time

Here we see the ball has constant velocity from points A to B, meaning zero acceleration. Then as the ball drops at point B we have an increase in velocity, meaning positive acceleration. But as the ball climbs toward point D, velocity slows down, thereby giving us negative acceleration.

Again, we have Time represented on one axis of the graph. Time, being scalar rather than a vector quality, will continue to move forward. Velocity, being a vector quality, can move either up or down. Because velocity is a vector quality, acceleration is also a vector quality.

As you can see, acceleration is positive when velocity is moving away from zero; acceleration is negative when velocity is moving closer to zero. Similarly, velocity is positive when displacement is moving further from zero, and negative when displacement is moving closer to zero.

But...why does the ball accelerate as it drops down from point B to point C? And why does it decelerate on its way from point C to point D? We'll look at this over the next few pages.

If you hold your arm out to your side and let go of your pencil, what happens? It falls to the floor, right? But did it have constant speed as it fell, or was it accelerating?

Galileo tested this question with an inclined plane to slow down the motion enough that he could study it. He determined that an object gains speed as it falls.

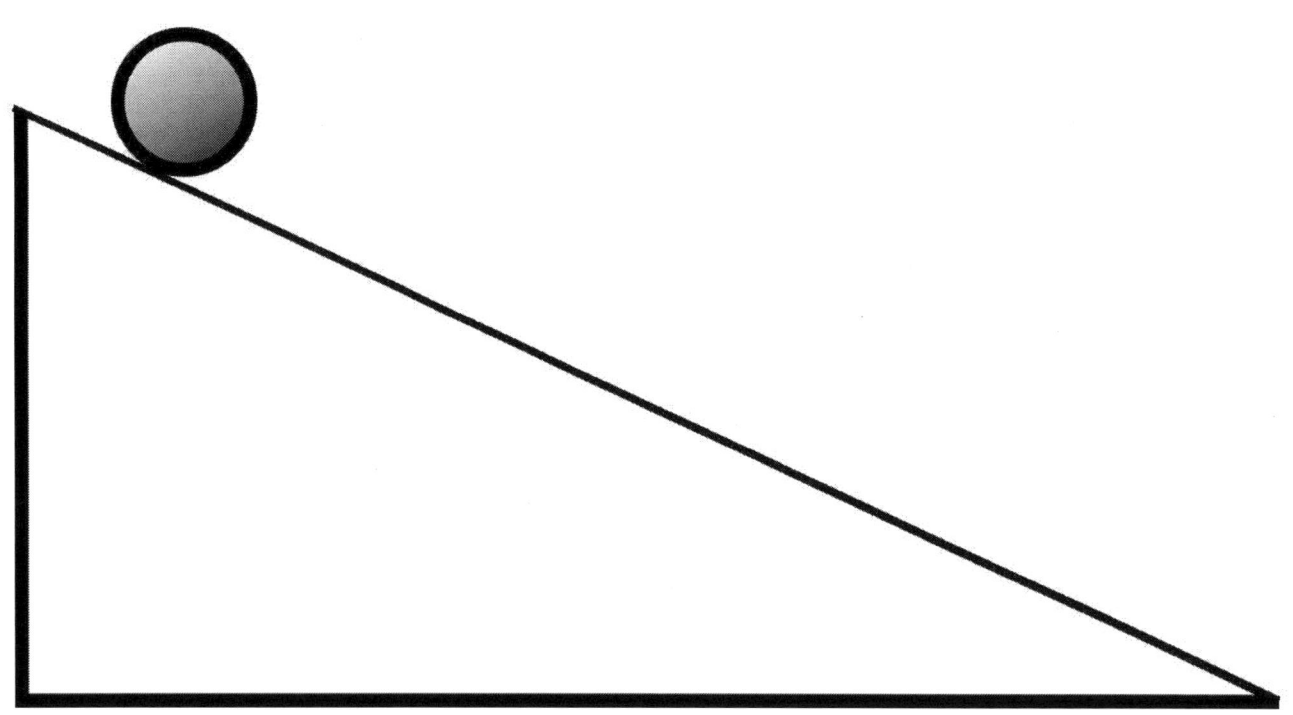

And not only does an object gain speed as it falls, but the distance it falls can be found with the formula

$$d = \frac{1}{2}at^2$$

CONTRIBUTORS TO PHYSICS

Galileo Galilei
1564-1642

Born in Italy, at age 22 Galileo was studying medicine at the University of Pisa when he attended a mathematics lecture and decided to pursue math and science instead. He went on to make many important contributions to these subjects.

When it comes to falling straight down, we've found that near Earth's surface, the acceleration of falling is $9.8 m/s^2$. This constant rate of acceleration is represented by a lowercase **g**. So for a falling object near Earth's surface, the way to calculate how far an object fell is:

$$d = \frac{1}{2} g t^2$$

and the way to calculate a falling object's velocity is:

$$v = gt$$

You'll notice in these equations, there is no variable to account for the object's weight or size. This is because, ignoring air resistance, all objects fall with the same rate of acceleration no matter their size.

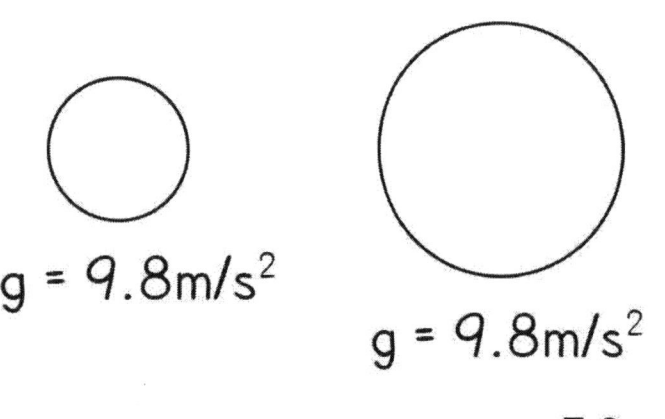

$g = 9.8 m/s^2$

$g = 9.8 m/s^2$

$g = 9.8 m/s^2$

This can be tested by anyone with two objects of the same size but different weight, for instance, a full bottle of shampoo and an empty bottle of shampoo. If both are dropped from a height of 100 meters, they'd hit the ground simultaneously after 4.51 seconds.

		time	velocity	distance
FULL	EMPTY			
		1s	9.8m/s	4.9m
		2s	19.6m/s	19.6m
		3s	29.4m/s	44.1m
		4s	39.2m/s	78.4m

With objects of differing sizes and shapes however, air resistance can slow the acceleration. Think of confetti thrown into the air. Each tiny piece of paper flitters about as it falls.

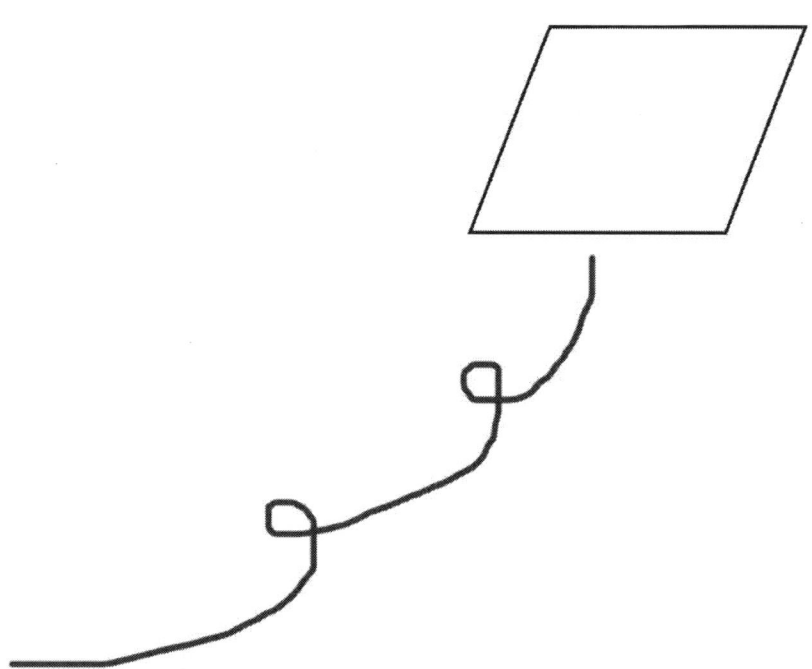

But, if placed into a vacuum (a container with all the air pumped out) where there is no air resistance, confetti would fall straight to the ground at the same rate as a bowling ball.

9.8m/s^2

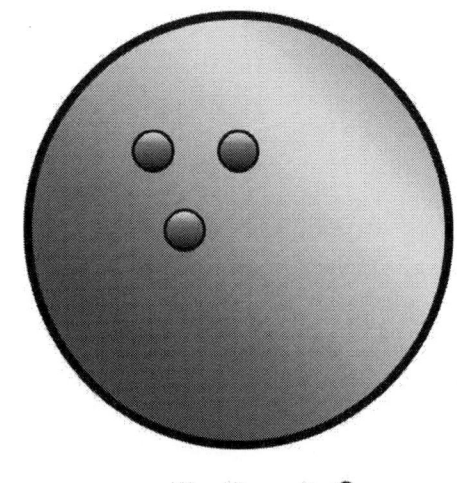

9.8m/s^2

Similarly, an object rolling on the ground is slowed by air resistance as well as a force called **friction**. This is why if a driver releases the gas pedal the car will eventually come to a stop.

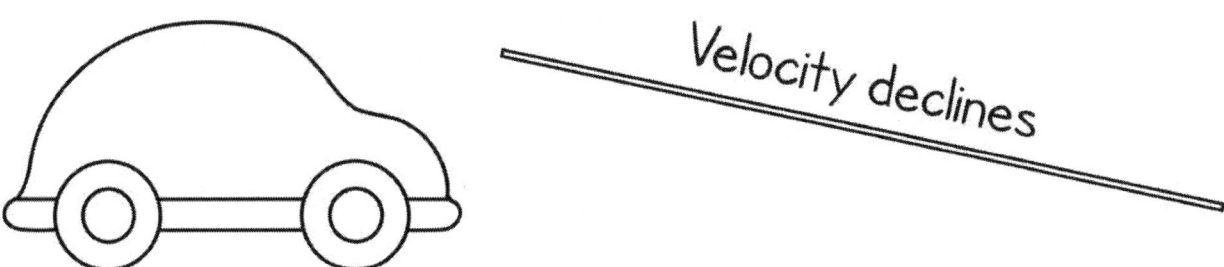

Velocity declines

But, absent friction and air resistance, that same car would continue on perpetually until it hit something or otherwise malfunctioned.

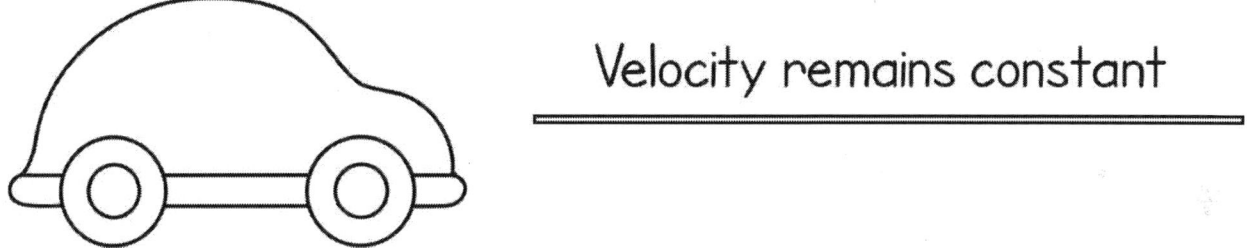

Velocity remains constant

Galileo realized this and made the claim that no force is needed to keep an object in motion. Isaac Newton rephrased this principle in his First Law of Motion:

An object at rest tends to stay at rest. An object in motion tends to stay in motion.

CONTRIBUTORS TO PHYSICS

Isaac Newton
1643-1727

Isaac Newton was born in England, a farmer's son, born premature and not expected to survive. His father died a few months after he was born and at age 12 his mother pulled him out of school to have him tend the farm, but he wasn't good at it and returned to school. It was his uncle who later convinced his mother to have him enroll at the University of Cambridge, and he went on to become an extremely influential physicist and mathematician.

When an object is in motion, it takes an outside force to change that. Friction and air resistance are two examples of such force. An object's tendency to resist changes in motion is **inertia**. The greater the object's mass, the more inertia it has.

greater inertia   less inertia

It takes more force to change the inertia of a semi-truck than to change the inertia of a bicycle.

This property leads us to **Newton's Second Law of Motion**:

$$F = ma$$

Where F is force, m is mass, and a is acceleration.

The more force there is on an object, the more it accelerates, but the more massive it is, the more it resists acceleration.

The standard unit for measuring force is the **newton**, represented by the letter **N**. To calculate in newtons, mass should be measured in kilograms (kg) and acceleration should be measured in meters per second squared (m/s^2). A newton is equal to approximately $1/4$ a pound of pressure.

If you have a box of books with a mass of 12.35kg and you push it across the room with acceleration of $0.75m/s^2$, how much force did you apply?

$$F = ma$$
$$F = (12.35kg)(0.75m/s^2)$$
$$F = 9.26N$$

What if you have a salt shaker with a mass of 25g you need to slide across the table to someone else?

First you have to convert grams to kilograms:

$$25g\left(\frac{1kg}{1000g}\right) = 0.025kg$$

Then, if you know you slid it across the table with a force of 0.056N, what was the salt shaker's acceleration?

$$a = F/m$$
$$a = 0.056N/0.025kg$$
$$a = 2.24m/s^2$$

What about if you know an object's acceleration and force but not its mass? Then you'd use:

$$m = F/a$$

If an object is accelerating at a rate of $4.7m/s^2$ with a force of 7.92N, what is its mass?

$$m = 7.92N/4.7m/s^2 = 1.685kg$$

What if it takes twice the force to accelerate a different object to $4.7m/s^2$?

$$m = 15.84N/4.7m/s^2 = 3.37kg$$

Then the object must have twice the mass!

Now try a couple questions on your own!

How much force does it take to move an object with a mass of 512kg at an acceleration of $1.2m/s^2$?

If a mass of 36.4kg is moving with a force of 17N, how fast is it accelerating?

If an object is accelerating at a rate of $0.82m/s^2$ by a force equalling 6.2N, what is its mass?

So, back to friction and air resistance, these forces naturally slow a moving object, which is why we don't see perpetual motion at play with everything around us. A wagon rolling along a sidewalk will eventually slow to a stop if there isn't enough force applied to keep it in motion, because friction and air resistance act as forces in the opposite direction. If there weren't any opposing forces, however, velocity would remain constant. The object in motion would stay in motion. But where there is matter, there is force. When an object isn't moving, force is at equillibrium.

According to **Newton's Third Law of Motion**:

When an object exerts force on another object, the second object exerts an equal but opposite force on the first.

Consider a house. It is stationary on the ground. Does it have force? Yes! Earth's gravity pulls down on the house, and the house pulls back with equal force.

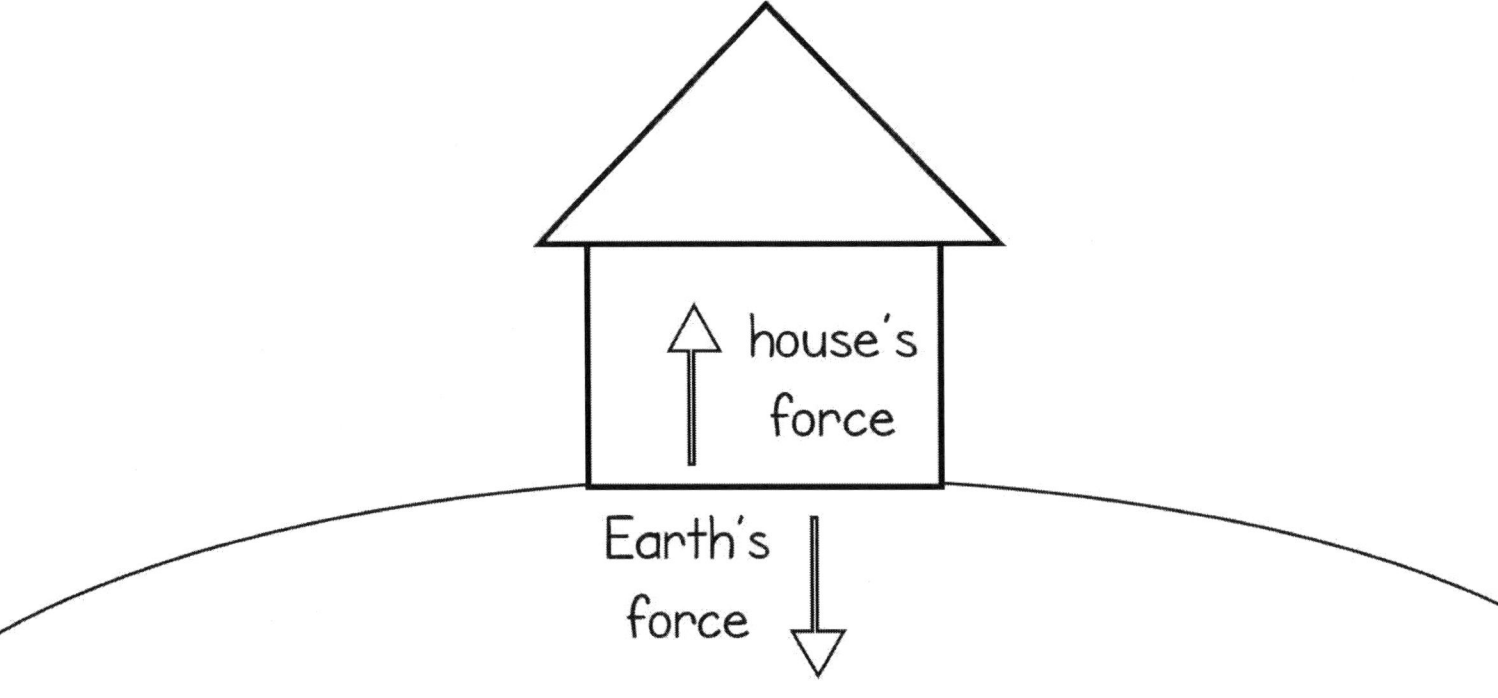

The formula to determine the force of gravity/attraction between any given objects is

$$F = G \frac{Mm}{r^2}$$

M stands for the larger object's mass
m stands for the smaller object's mass.
r is the distance between the two objects
G is the universal gravitational constant

This formula can be used to determine the amount of attraction between any two objects, including Earth and the moon or other planets.

Just like Earth's gravity keeps a house stationary on the ground, Earth's gravity also pulls on the moon, keeping it in elliptical orbit around the Earth instead of continuing off into space in a straight-line motion.

The moon likewise pulls on the Earth with an equal amount of force, resulting in various phenomena on Earth such as the ebb and flow of the tides.

Every second, the moon moves approximately one millimeter away from the straight-line path it would naturally take if Earth's gravity wasn't pulling on it.

Tides happen because the water closest to the moon is pulled more strongly by the moon's gravitational pull than the Earth is, and the Earth is pulled more strongly by the moon than the water furthest from the moon on the opposite side of the Earth.

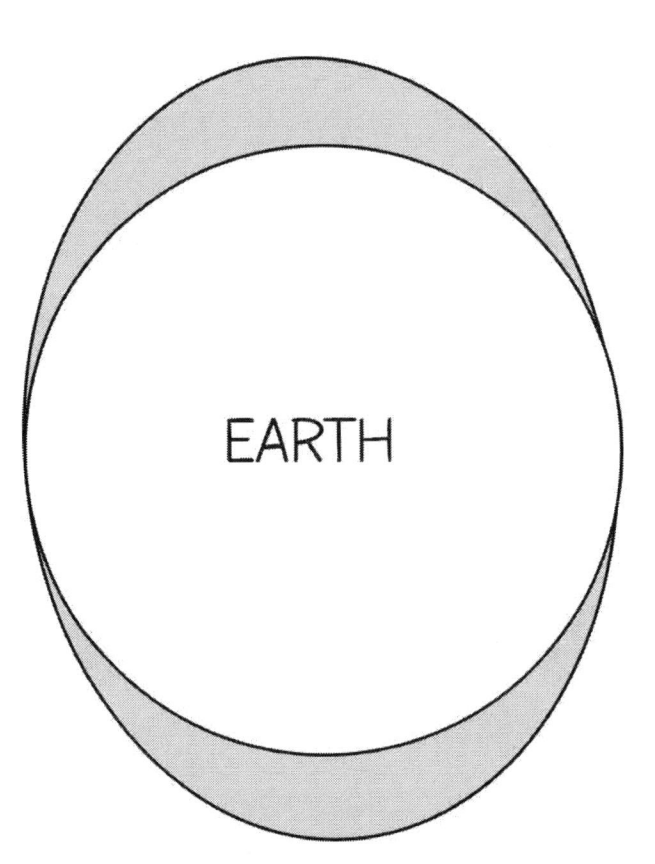

EARTH

So far we've learned two formulas for finding force.

$$F = ma$$

and

$$F = G\frac{Mm}{r^2}$$

But when it comes to gravity, we also have

$$F = W$$

That's right: an object's weight is equal to the amount of force exerted on it by gravity.

Think about it: if you put an object on a scale, the weight registered has everything to do with the Earth's gravitational pull on that object.

11.96 lb

Close to the Earth's surface, the acceleration associated with falling is **g**, 9.8m/s^2. And why do objects fall to the ground? Because of the force exerted on them by the Earth's gravitational pull. So if

$$F = ma$$

and

$$a = g$$

then when it comes to gravity and falling,

$$F = mg$$

The next question to ask is, does acceleration **g** go away when an object hits the ground? No! Force is still there, but because the object maintains an equal and opposite force, it balances out to zero.

So, what happens on a scale?

The object on the scale is being pulled toward the Earth with acceleration **g**. At the same time, the object on the scale is pulling on the Earth with the same amount of force.

However, the object and the scale are also exerting force on each other.

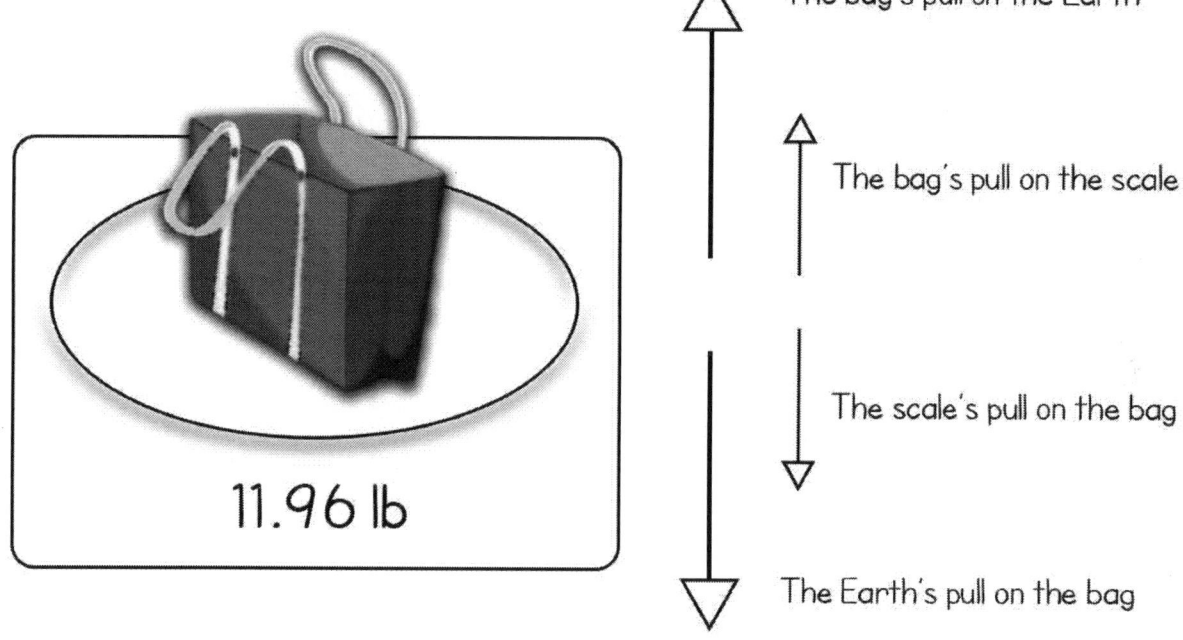

What happens is, the object on the scale has two forces pulling down on it: the force exerted on it by the scale, and the force exerted on it by the Earth.

The force shared between the object and the scale cancel out to zero, and the force between the scale and the Earth cancel out to zero, so what is it the scale measures? It measures the force exerted on the object by the pull of the Earth.

Therefore:

$$F = W$$

and

$$W = mg$$

Once we realize that **g** plays a role in *determining* weight, it is easier to understand why objects fall to the ground with the same acceleration regardless of weight.

But what about mass?

Well, we know that

$$F = mg$$

and

$$F = G\frac{Mm}{r^2}$$

So we can now reason that:

$$mg = G\frac{Mm}{r^2}$$

The mass of the smaller object cancels out on each side, leaving us with:

$$g = G\frac{M}{r^2}$$

This shows us that not only does an object's weight not change its acceleration when dropped near the surface of the Earth, but neither does its mass.

The only mass that matters here in determining **g** is the mass of the Earth, **M**. The mass of the smaller object on the Earth is irrelevant to the rate of **g**.

This proves what we learned earlier that, absent other forces slowing their acceleration, objects close to the Earth's surface will fall to the Earth at a rate of $9.8 m/s^2$ regardless of their size.

Now, let's test our equation to prove the rate of g

M is the mass of the Earth. Scientists have calculated the Earth's mass to be approximately 6×10^{24} kg

Since we only have one mass in this equation, that of Earth, r will be Earth's radius. The surface of the Earth is 6.4×10^6 meters from Earth's center.

And G, the universal gravitational constant, has a value of 6.67×10^{-11} Nm^2/kg^2

So let's put these numbers into the equation.

$$g = G \frac{M}{r^2}$$

$$g = 6.67 \times 10^{-11} \frac{N\,m^2}{kg^2} \left(\frac{6 \times 10^{24}\,kg}{(6.4 \times 10^6\,m)^2} \right)$$

$$g = 9.8 \text{ N/kg}$$

Since a newton is measured in kilograms times meters per second squared, N/kg = m/s^2

$$g = 9.8 \text{ m/s}^2$$

So, now that we've adequately demonstrated why neither the weight nor mass of an object have an impact on *g*, what is the difference between weight and mass, anyway?

Weight is, technically speaking, the amount of gravitational pull on the object. The further an object is from Earth's gravitational pull, the less weight it has. We can see this because in our equation, as an object's distance from Earth increases, *r* increases, and as *r* increases, the force exerted by Earth decreases.

Mass is the amount of matter that makes up an object. This quantity will remain constant regardless of location in regard to Earth or any other object.

If you have an apple, the amount of matter making up that apple doesn't change if you move it from one place to the next. But the amount of gravitational pull on it can change based on where it is. If you put the apple on the moon, it would still have the same amount of mass, but it would have less weight than it does on Earth.

Review!

These are the equations you've learned:

$$F = ma \quad a = F/m$$
$$m = F/a$$

$$F = mg \quad W = mg$$
$$F = W$$

$$F = G\frac{Mm}{r^2}$$
$$g = G\frac{M}{r^2}$$

And these are some of the concepts you've learned:

Objects gain speed as they fall. Neither mass nor weight affect this rate of acceleration.

An object's tendency to resist changes in motion is called **inertia**. The greater the object's mass, the more inertia it has.

Newton's Three Laws of Motion:

1. An object in motion tends to stay in motion.
2. The more force there is on an object, the more it accelerates, but the more massive it is, the more it resists acceleration.
3. When one object exerts force on a second object, the second object exerts an equal but opposite force on the first object.

Draw arrows to show the forces between this box and the ground. If the Earth's force on this box is 14.2 N, what is the box's mass?

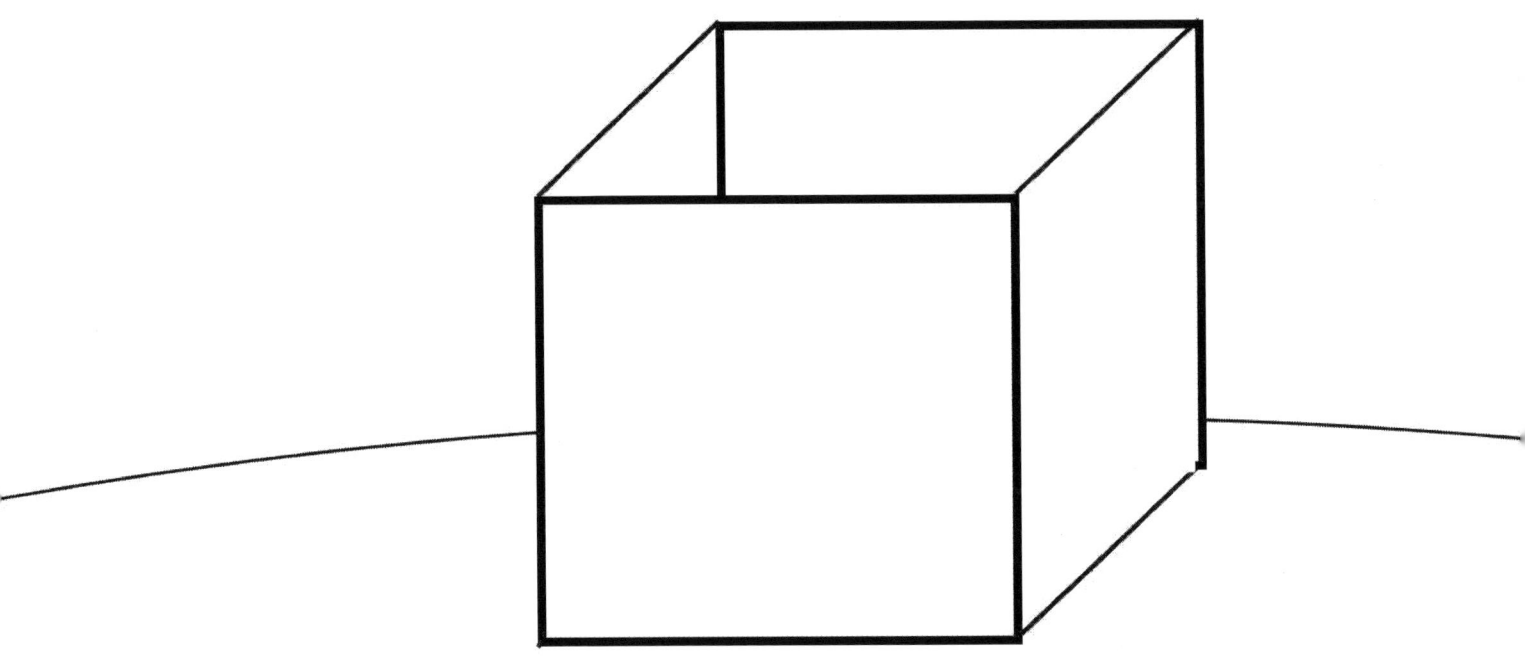

Add to the diagram to show a representation of how the moon's gravitational pull affect's the Earth's water. Draw a dotted line to show how the moon would move if it was not affected by Earth's gravity.

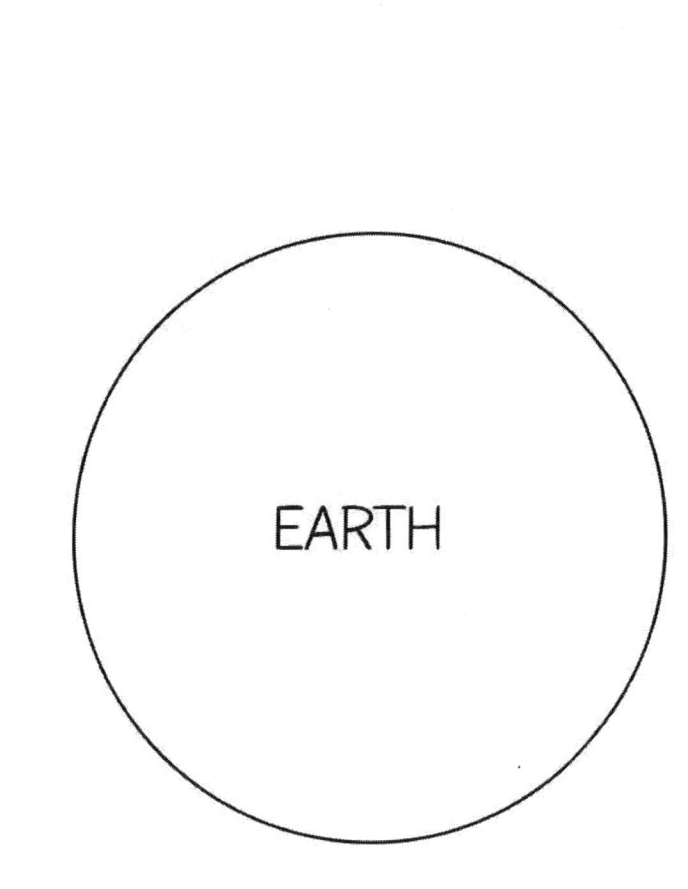

EARTH

Here you see a ceramic cat statue on a scale. Draw and label arrows to show the different forces at play here. Remember there are forces between the cat and the scale, the scale and the Earth, and the cat and the Earth. Circle the arrow that shows which force is calculated by the scale.

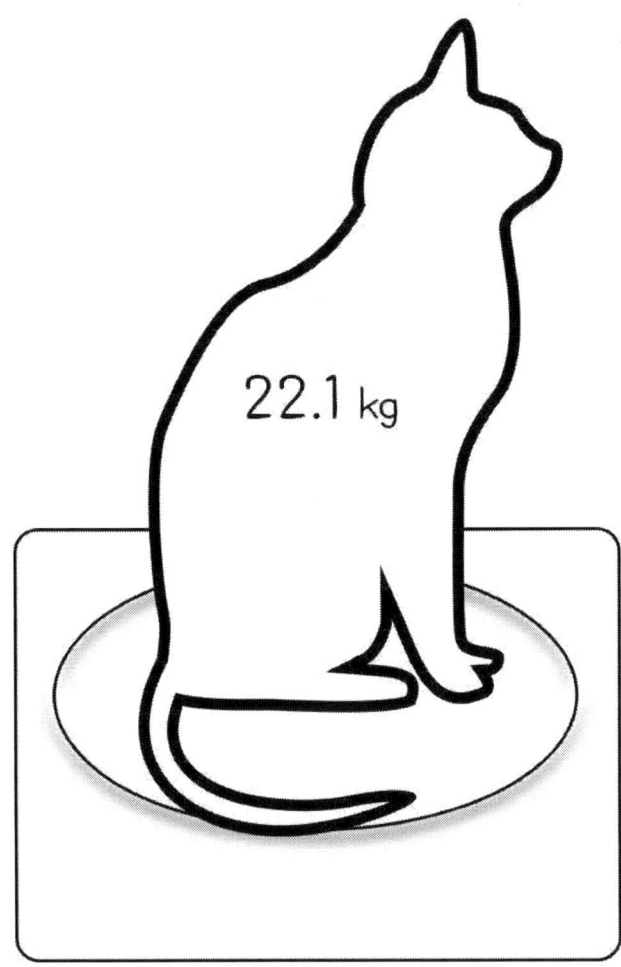

22.1 kg

Using the statue's mass and the constant **g**, calculate its force.

So far when learning about *g* we've been looking at examples of objects with a straight drop to the ground. Now we'll look at what happens when an object also has forward motion while it falls.

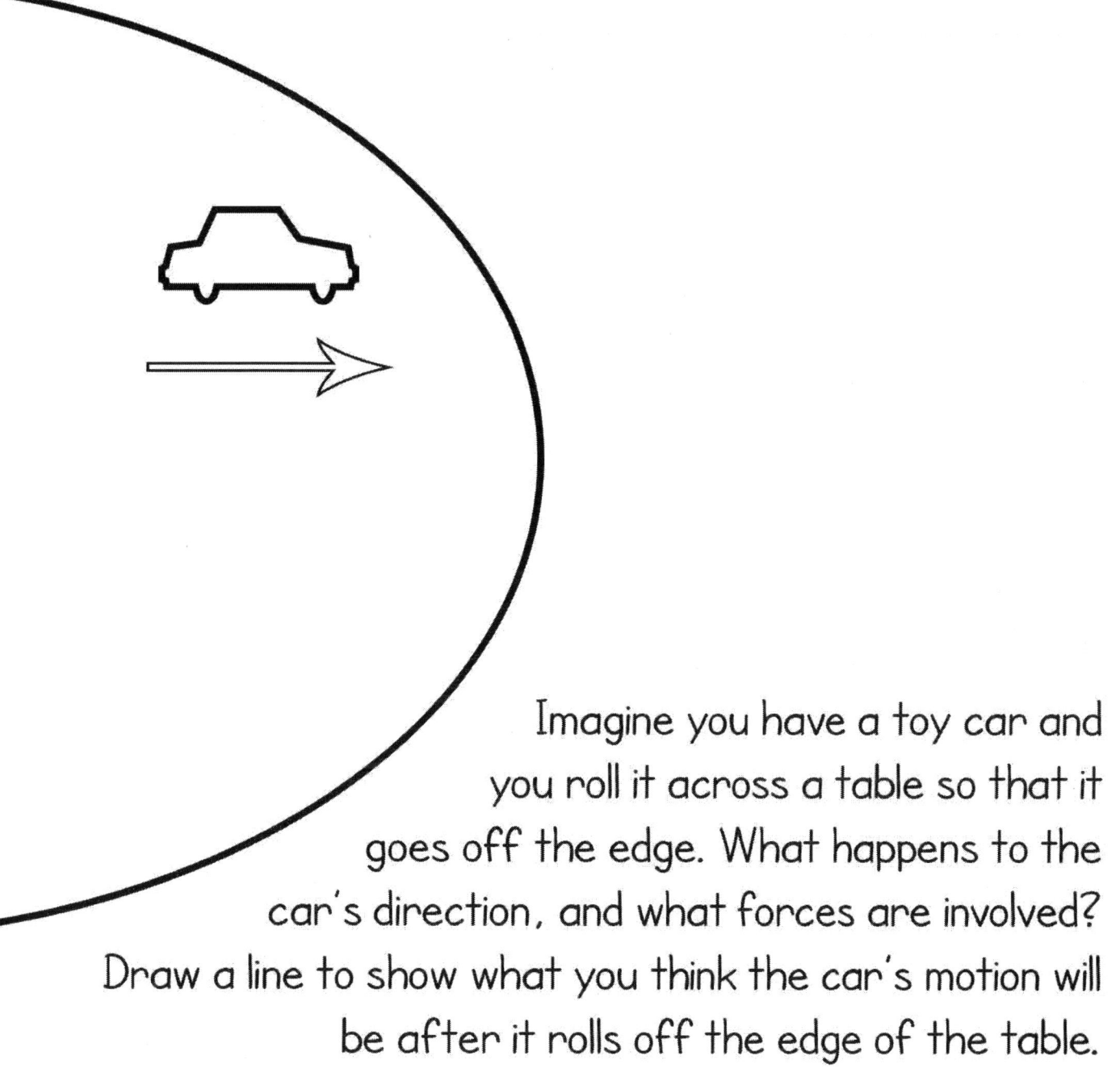

Imagine you have a toy car and you roll it across a table so that it goes off the edge. What happens to the car's direction, and what forces are involved? Draw a line to show what you think the car's motion will be after it rolls off the edge of the table.

When the car rolls off the table, there is more than one force in motion. There is the force of the car's momentum as it moves forward, and there is the force of gravity pulling it down.

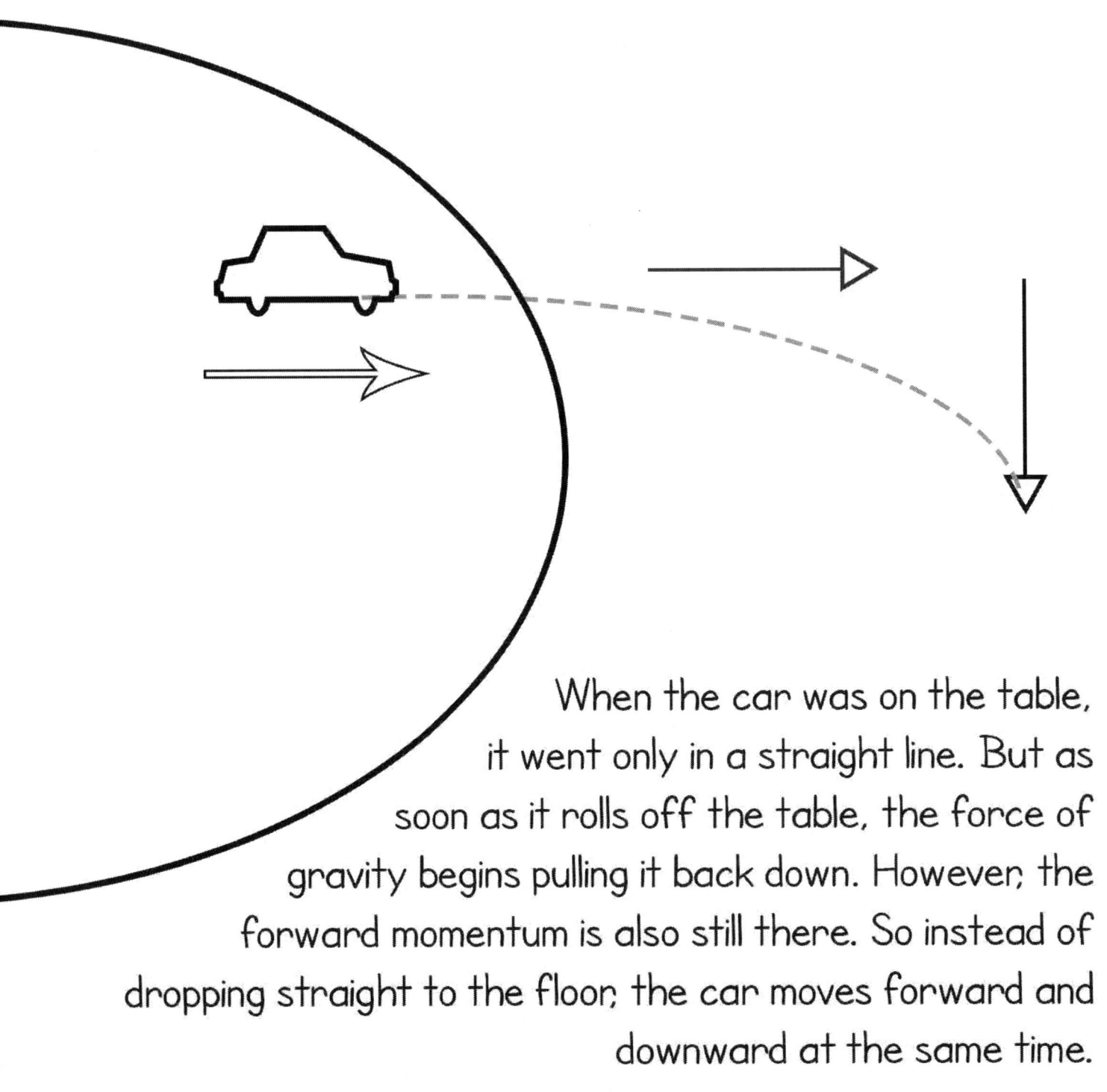

When the car was on the table, it went only in a straight line. But as soon as it rolls off the table, the force of gravity begins pulling it back down. However, the forward momentum is also still there. So instead of dropping straight to the floor, the car moves forward and downward at the same time.

So, now we ask, does $t = \sqrt{\dfrac{2d}{g}}$ still hold true if the object is also moving forward while it's falling?

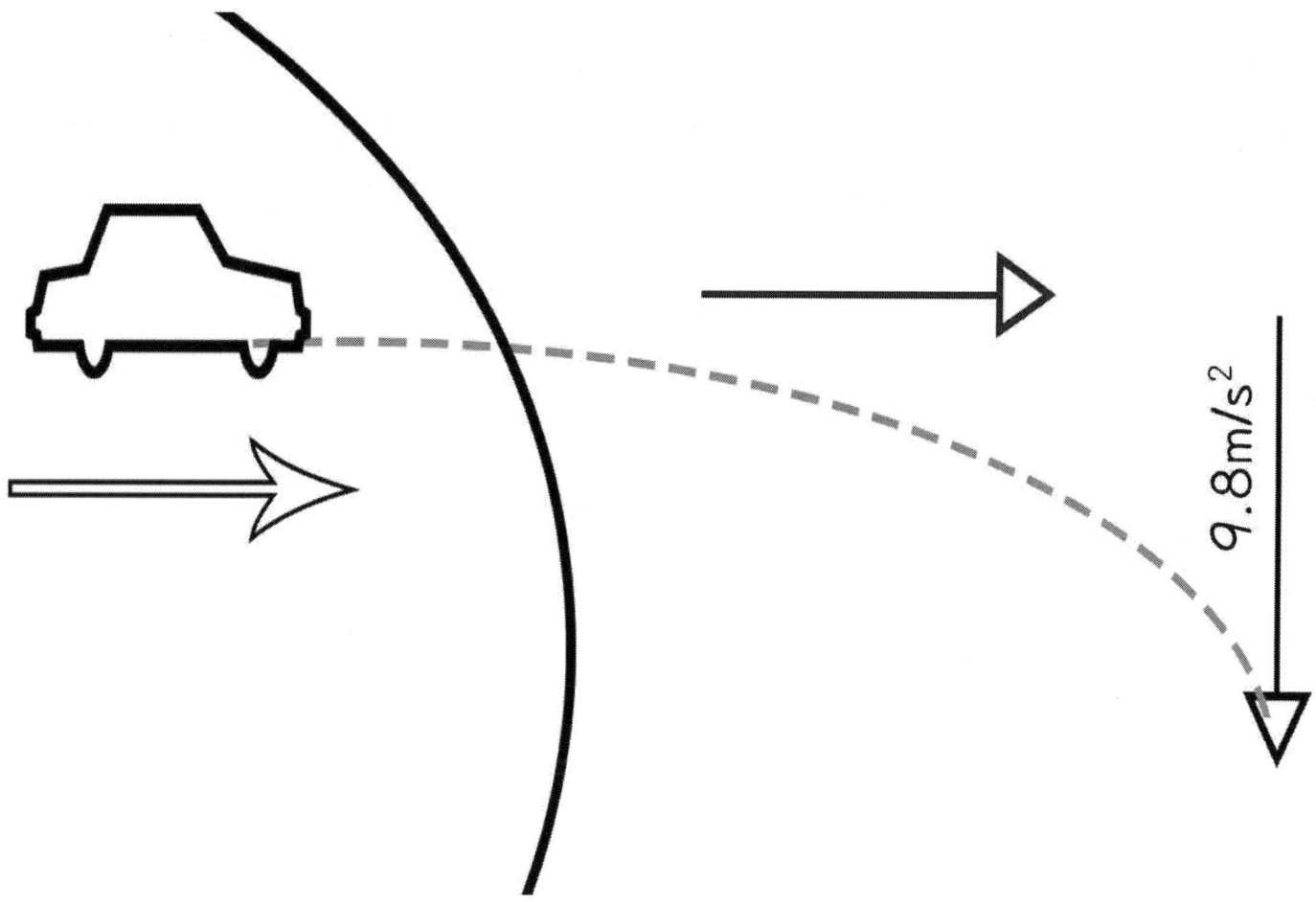

9.8m/s²

The answer is yes! The forward momentum doesn't have any effect on the rate of the downward pull. *g* remains constant.

Let's say the table is 0.87m tall. How long does it take for the car to hit the ground?

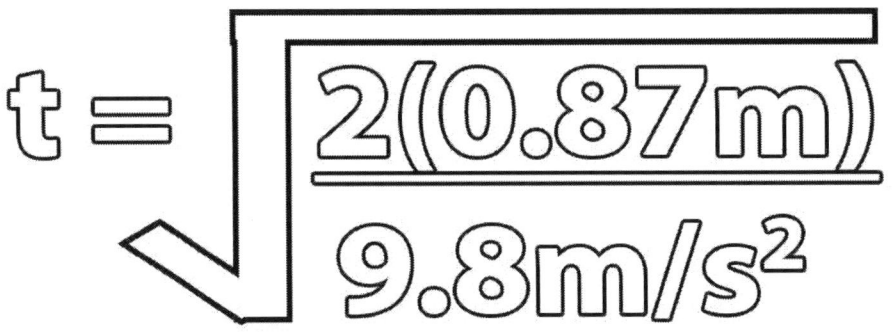

$$t = 0.42s$$

It takes 0.42 seconds for the car to reach the ground. Because the force of gravity works independently from the force moving the car forward, the only distance taken into account is the straight up-and-down distance from the surface of the table to the floor. The distance the car moves forward while falling doesn't have any impact on how long it takes the car to fall.

Now, let's say that at the same time that the toy car leaves the surface of the table, someone else drops a different toy car from the same height. Which car reaches the ground first?

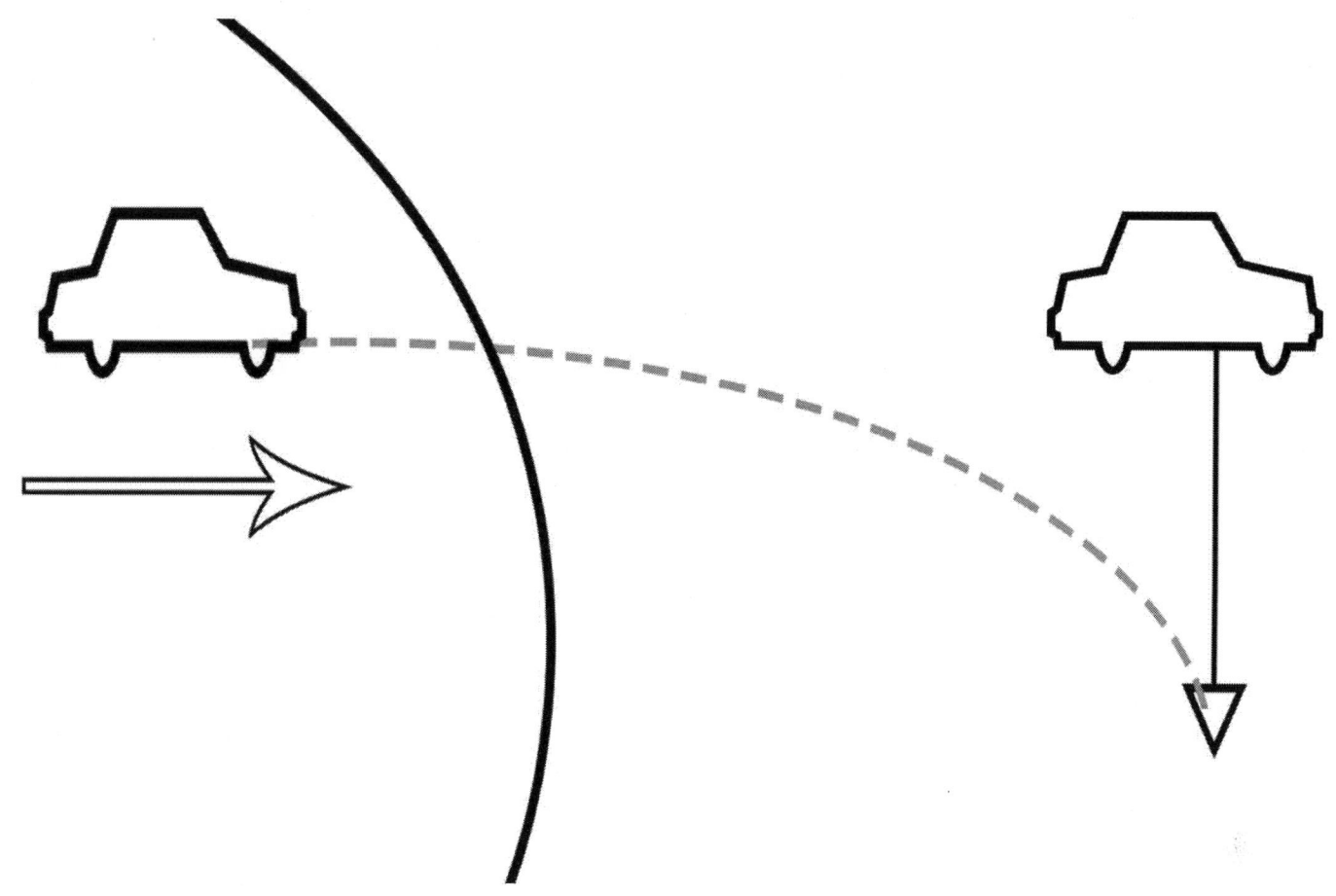

If you said they reach the ground at the same time, you are correct! Again, the forward motion doesn't change the rate of falling.

We've analyzed the vertical force on the car from the Earth's gravity, but what about the forces at play horizontally?

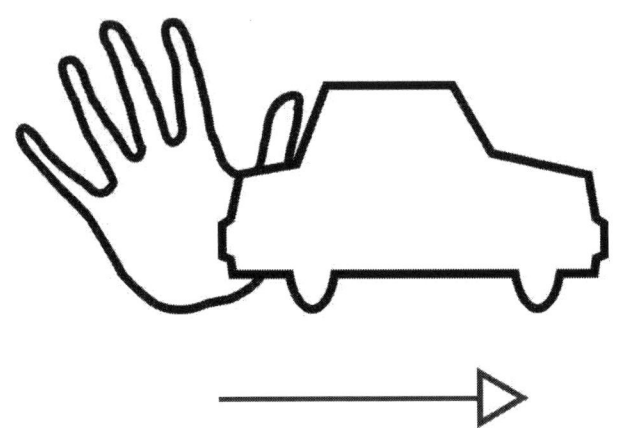

When you push on the car with your hand, you're applying the force which moves it forward. Of course, we remember from Newton's Third Law of Motion that if you apply force to the car, an equal and opposite force is exerted by the car. Where does this force go?

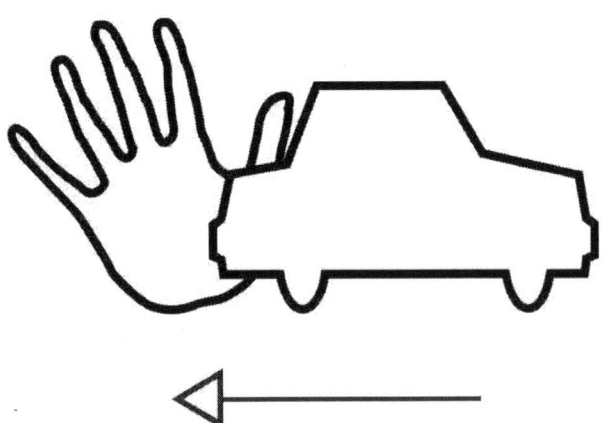

The equal and opposite force goes from the toy car to your hand. These forces are equal and opposite, so they cancel each other out. So why does the car move?

The car moves forward because the force applied to it is greater than the other forces which are keeping it in place. Namely, the friction between the car and the surface of the table.

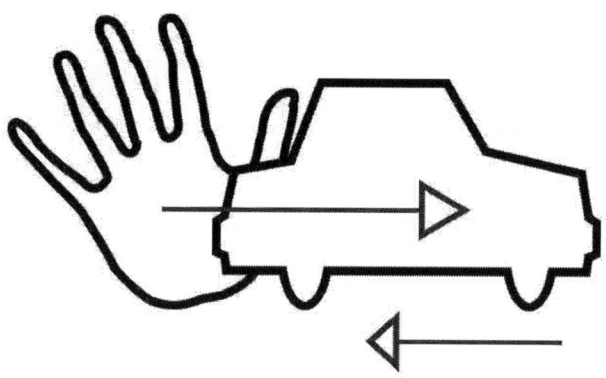

As long as you apply force greater than that of the friction between the car and the table (the opposing force), the car will accelerate.

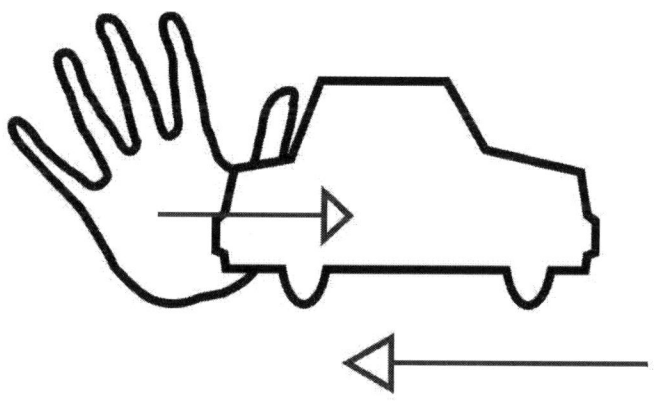

If the friction between the car and the table is greater than the force you apply, the car will not move forward.

And what about the opposing force the car puts on you? Remember, the opposing force from the toy car is equal to the force you exerted on it.

So, what will determine the effect of this force on your hand? Again, it has to do with whether this force is greater than, equal to, or less than the forces holding you and your hand in place.

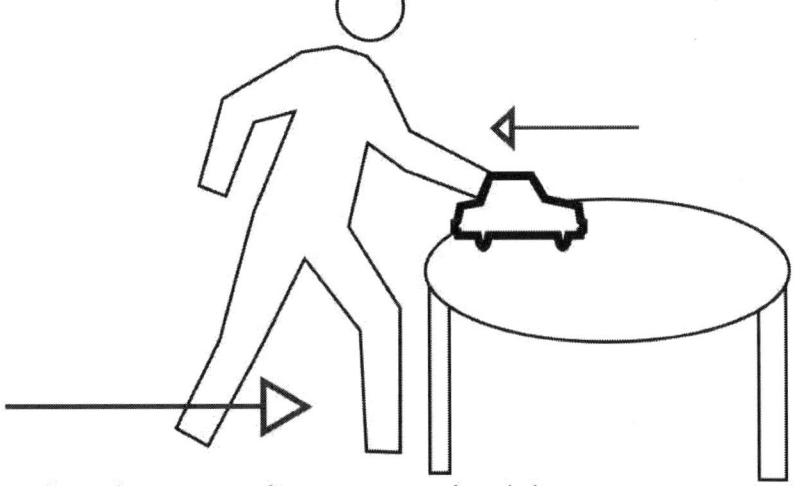

The muscles in your body use force to hold you up, and you also have friction with the ground. If the opposing force from the car is equal to or less than the forces which are keeping you where you're standing, the force from the car will not push you back.

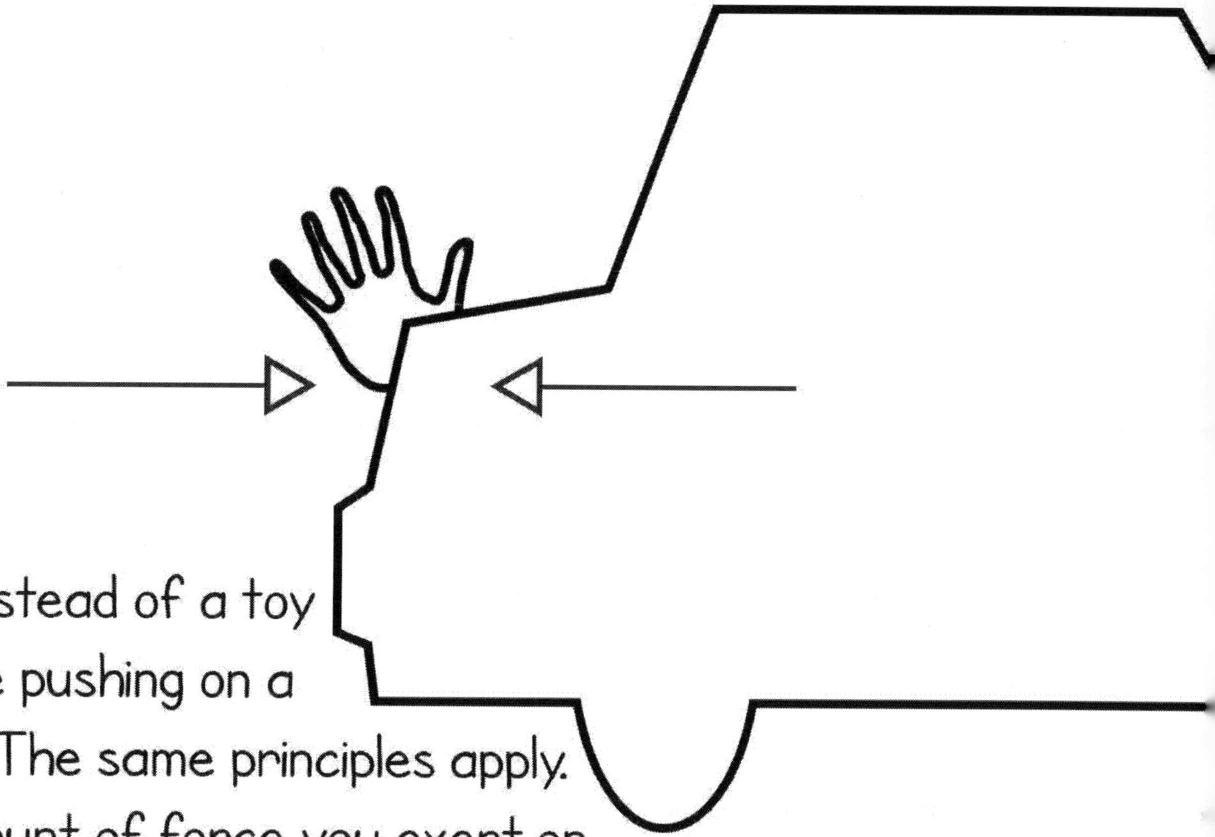

What if instead of a toy car you're pushing on a real car? The same principles apply. If the amount of force you exert on the car is less than the amount of friction the car has with the ground, the car will stay in place. If the amount of force you exert on the car is also less than the amount of force holding you upright, you also will stay in place.

But, if the amount of force you exert on the car is less than the amount of force holding you in place, the opposing force from the car will push you back.

So, let's go back to the toy car on the table. We've determined that if the table is 0.87 meters tall, it will take the car 0.42 seconds to fall to the ground. How far forward does it travel in this time?

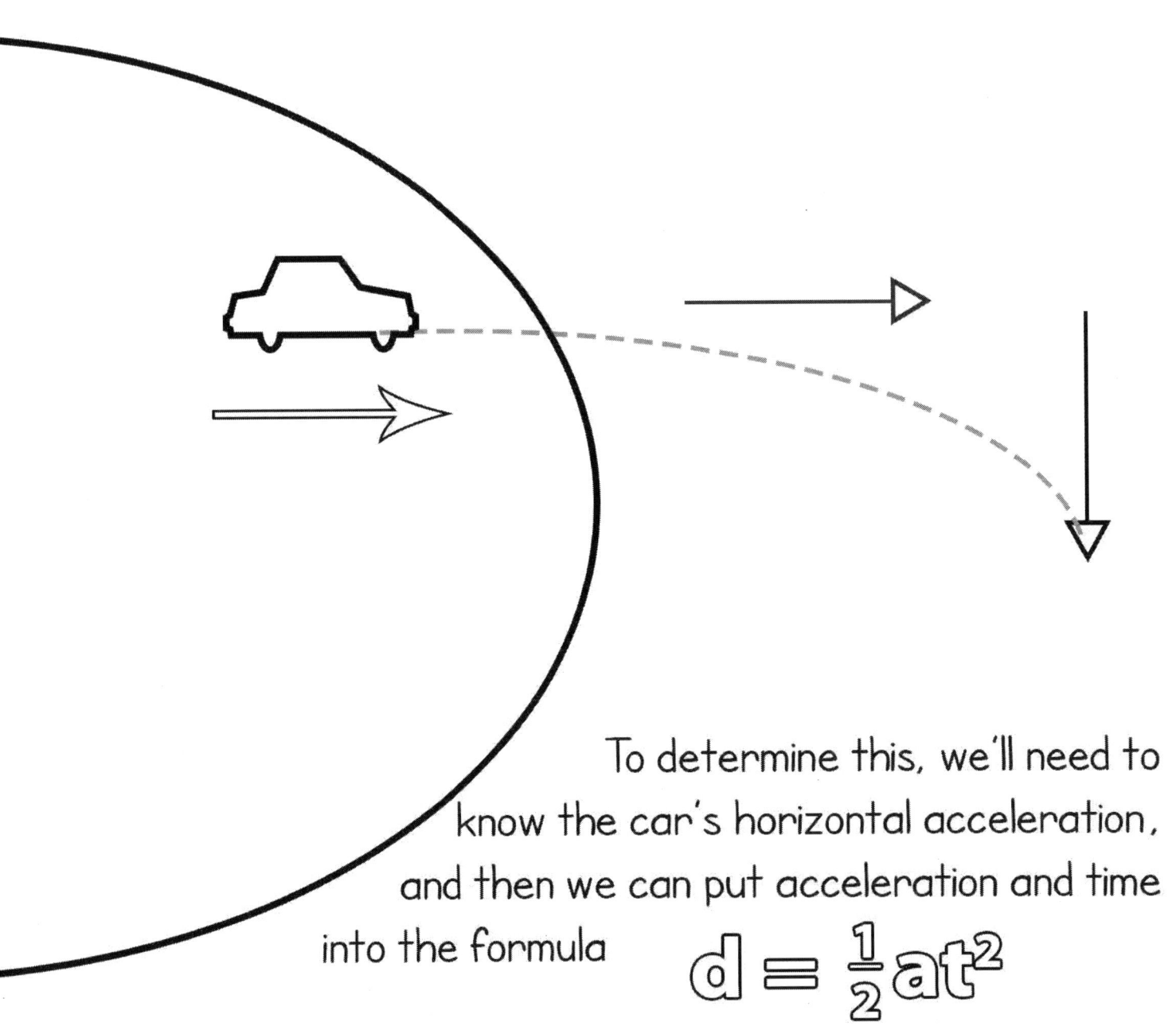

To determine this, we'll need to know the car's horizontal acceleration, and then we can put acceleration and time into the formula

$$d = \tfrac{1}{2}at^2$$

To determine the car's horizontal acceleration, we first need to divide the amount of force placed on the car by its mass.

Let's say the toy car has a mass of 0.25kg, and you apply 3 newtons of force to it as it sits on the very edge of the table.

$$a = F/m$$

$$a = 3N/0.25kg$$

$$a = 15m/s^2$$

Its horizontal acceleration is 15m/s².

Once it leaves the table, immediately after you applied the force to it, it only travels for 0.42 seconds before hitting the ground. How far forward does it travel in this time?

$$d = \frac{1}{2}at^2$$

$$d = \frac{1}{2}(15m/s^2)(0.42s)^2$$

$$d = 2.64m$$

In 0.42 seconds, the toy car travels 2.64 meters horizontally and 0.87 meters down.

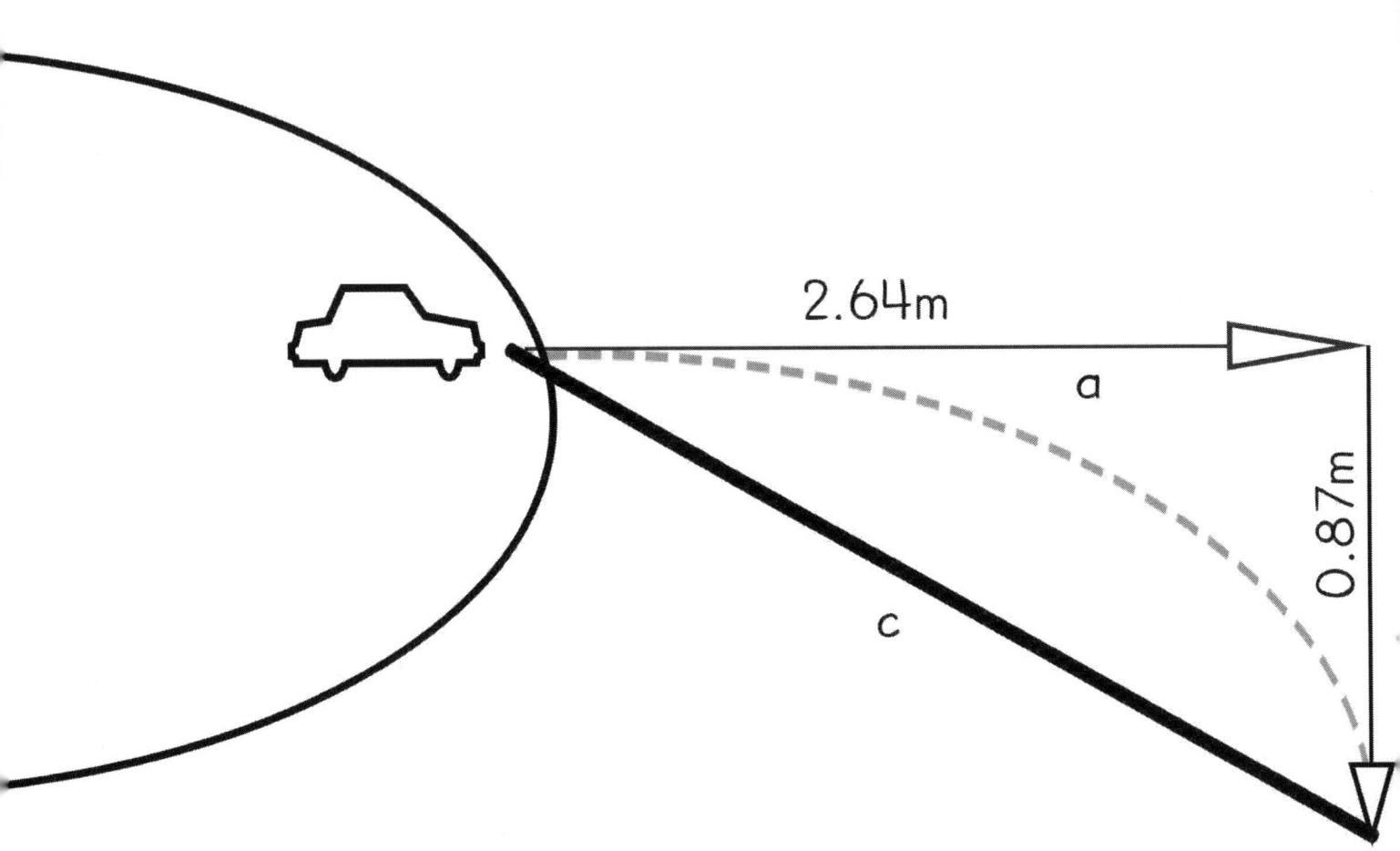

If the car travels 2.64 meters horizontally and 0.87 meters down to the floor, what is the car's displacement from its starting point?

For this we use some simple geometry. We know this is a right triangle because the horizontal forces and vertical forces are perpendicular to each other. So to find the length of the longest side, which is the car's displacement, we use $a^2 + b^2 = c^2$ and then take the square root of c^2 to find the length of c.

After doing the math, we find that in terms of displacement, the car traveled a total of 2.78 meters from where it began.

Now consider what you've learned about Isaac Newton's Third Law of Motion. If every force exerted receives an equal and opposite force in response, the total force involved balances out to zero. Put numerically, if you have 5 newtons of force in one direction and 5 newtons of force in the opposite direction, it balances out to a net force of 0 newtons. No force is exerted without an equal and opposing force meeting it, so the total amount of force doesn't change.

Another quantity that behaves similarly is energy.

Energy can be neither destroyed nor created, only transferred. Similar to how every force is met with an equal and opposite force, balancing out the total amount of force to zero, every time energy is spent by one object it is gained by another. The total energy at the end of any event is equal to the total energy before the event. This is the law of conservation of energy.

While force is measured in newtons, energy is measured in joules. One joule is the ability to exert a force of one newton over a distance of one meter.

1 joule = 1 newton(1 meter)

Energy is the capacity to do work. Energy released does work, and anything being worked on gains that energy. Energy becomes work, therefore, in physics, work and energy are equivalent concepts.

$$E = W$$

Be careful not to confuse W representing work with W representing weight. Choose a different color for each to help yourself distinguish between the two.

Work is defined as force times the displacement through which force is exerted.

$$W = Fd$$

When calculating work, the only force considered is that in the direction of movement. The equal and opposite force is ignored for the sake of calculating work.

You'll recall that an object's weight is equal to the force exerted on it by Earth's gravity.

$$W = mg = F$$

Therefore, we can refer to an object's weight in newtons for use in this formula. If you lift a book weighing two newtons' worth of force to a height of 1.4 meters, how much work did you do?

$$W = Fd$$
$$W = (2N)(1.4m)$$
$$W = 2.8 \text{ joules}$$

And if you lift that same book to a height of 2.8 meters, how much work did you do?

$$W = 2N(2.8m)$$
$$W = 5.6 \text{ joules}$$

Displacement doubled, so therefore work doubled.

Likewise if weight doubled, work would double.

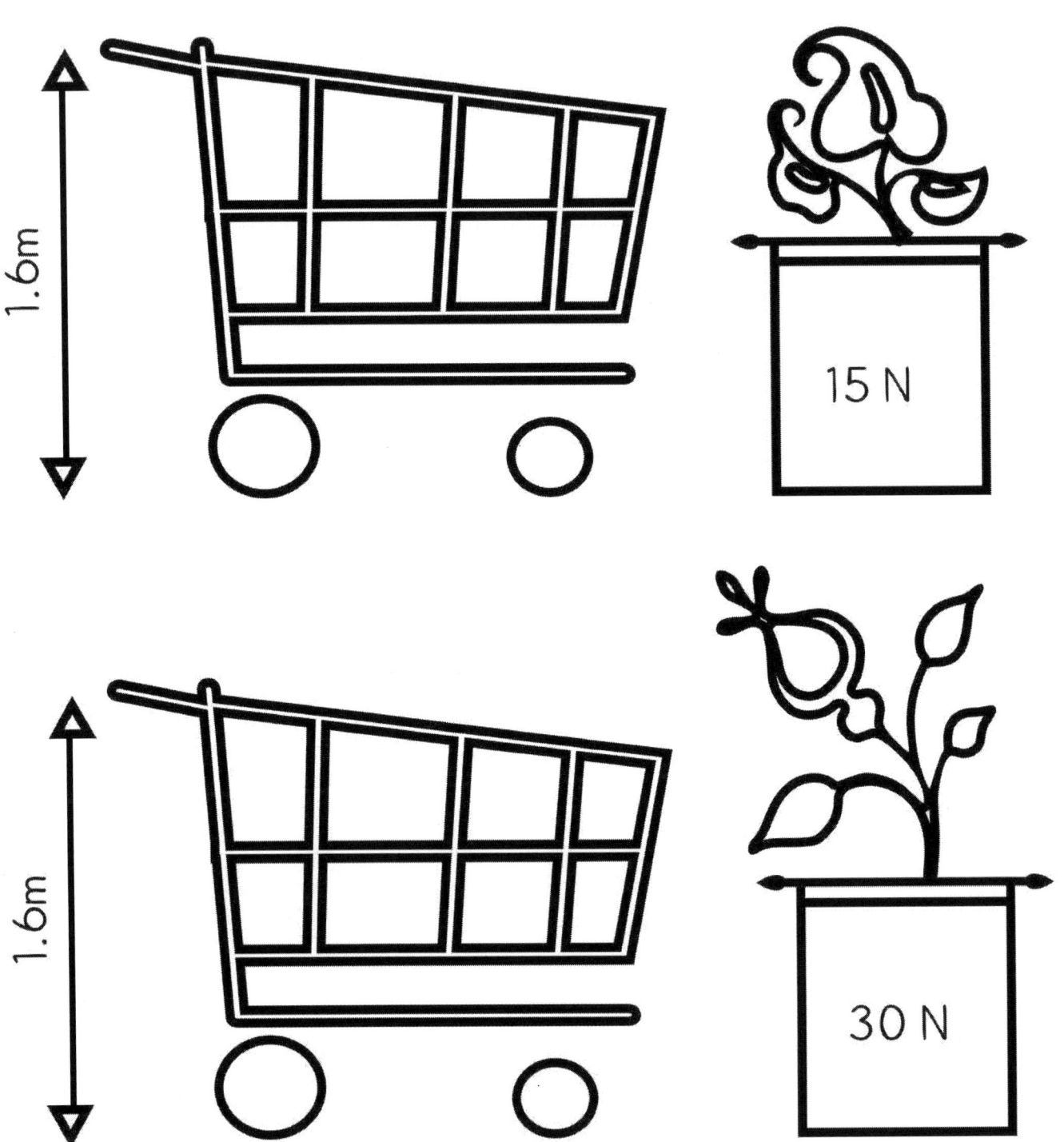

Two people go to the store and each purchase a large potted plant. Each plant is labeled with the amount of force it takes to lift it.

The plants must be lifted 1.6 meters to be placed in the cart. How much work is done in lifting each plant?

15 N

30 N

The 15 N plant needs to be pushed 12 meters to the checkout line. How much work is done?

The 30 N plant needs to be pushed 6 meters. How much work is done?

We said on page 87 that energy is neither created nor destroyed, but transferred from one source to another. When you lift a potted plant, you are using energy. This energy is called Kinetic Energy (K.E.), the energy of motion. But where does that energy go once the plant has been lifted?

The energy goes into the plant as Potential Energy (P.E.). If you lift a plant 2 meters, it has the potential now to fall 2 meters. As long as the plant remains 2 meters above the ground it has this Potential Energy.

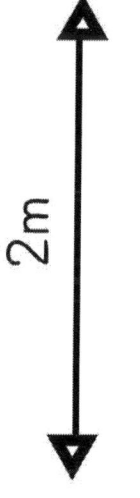

2m

If the plant were to drop to the ground again, the Potential Energy would gradually turn back into Kinetic Energy as it fell.

The formula for finding Kinetic Energy is

$$K.E. = \frac{1}{2}mv^2$$

We know this by following a logical progression with previously known formulas.

$$W = K.E.$$
$$W = Fd$$

$$K.E. = Fd$$
$$F = ma$$

$$K.E = mad$$
$$d = \frac{1}{2}at^2$$

$$K.E. = \frac{1}{2} ma^2t^2$$

$$K.E. = \frac{1}{2} m(at)^2$$

$$at = v$$

Therefore

$$K.E = \frac{1}{2} mv^2$$

As you can see, Work and Energy remain equivalent concepts.

$$K.E. = W = Fd = \frac{1}{2} mv^2$$

Using this formula, tell how much kinetic energy a 10,000kg sedan has when driving 60km/hr. Remember to convert km/hr to m/s in your calculating.

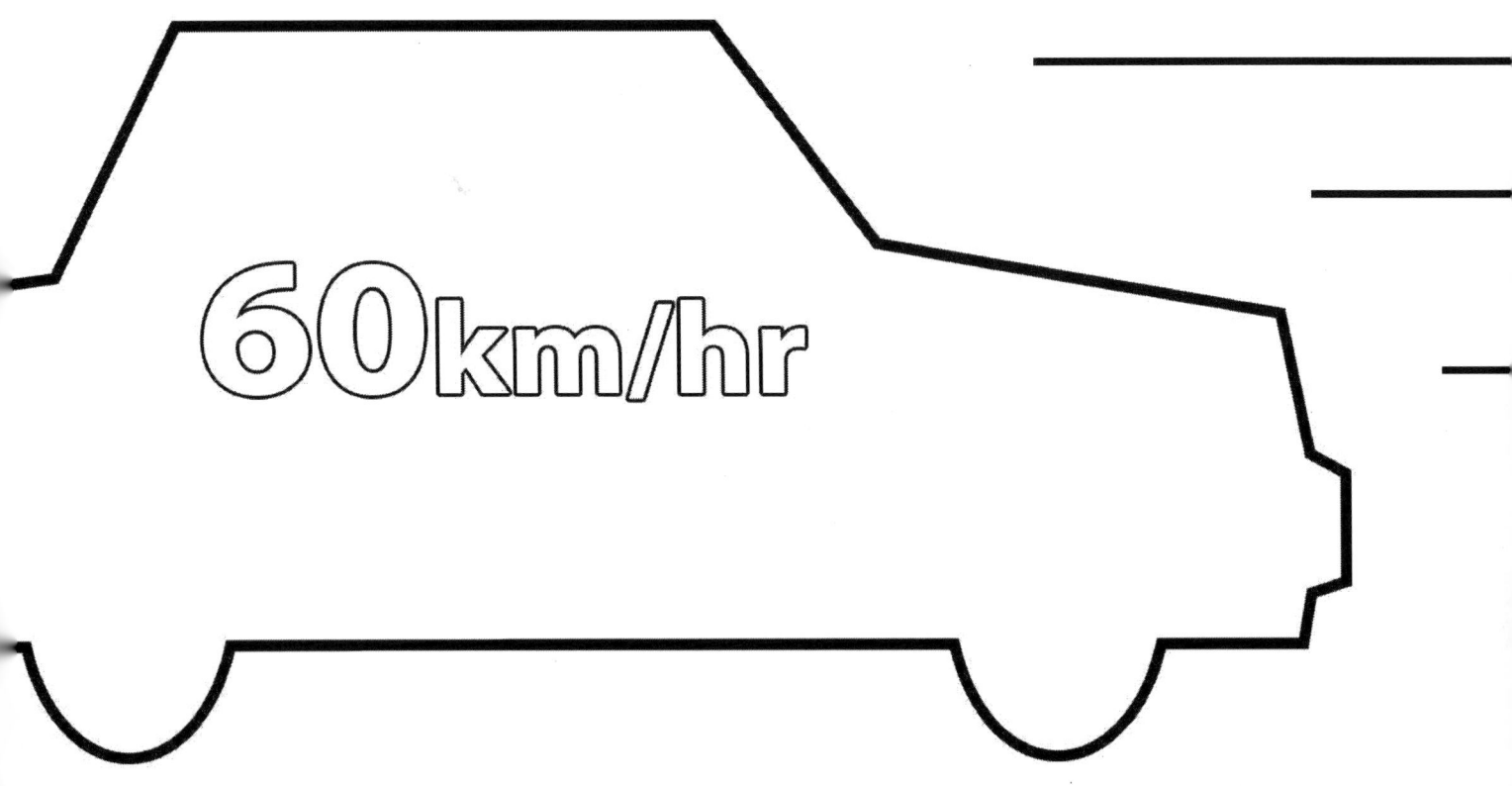

The formula for finding Potential Energy is

$$P.E. = mgh$$

Potential energy is the weight of the object times the height it's raised off the ground.

Since weight is mass times **g**

$$W = mg$$

Weight times height is **mgh**

$$P.E. = Wh = mgh$$

Since Kinetic Energy becomes Potential Energy, and when Potential Energy is released it becomes Kinetic energy again, we know that

$$K.E. = P.E.$$
$$\frac{1}{2}mv^2 = mgh$$

Imagine the same 10,000kg sedan on a lift at the mechanic. The car is lifted 1.6 meters off the ground. What is the car's potential energy?

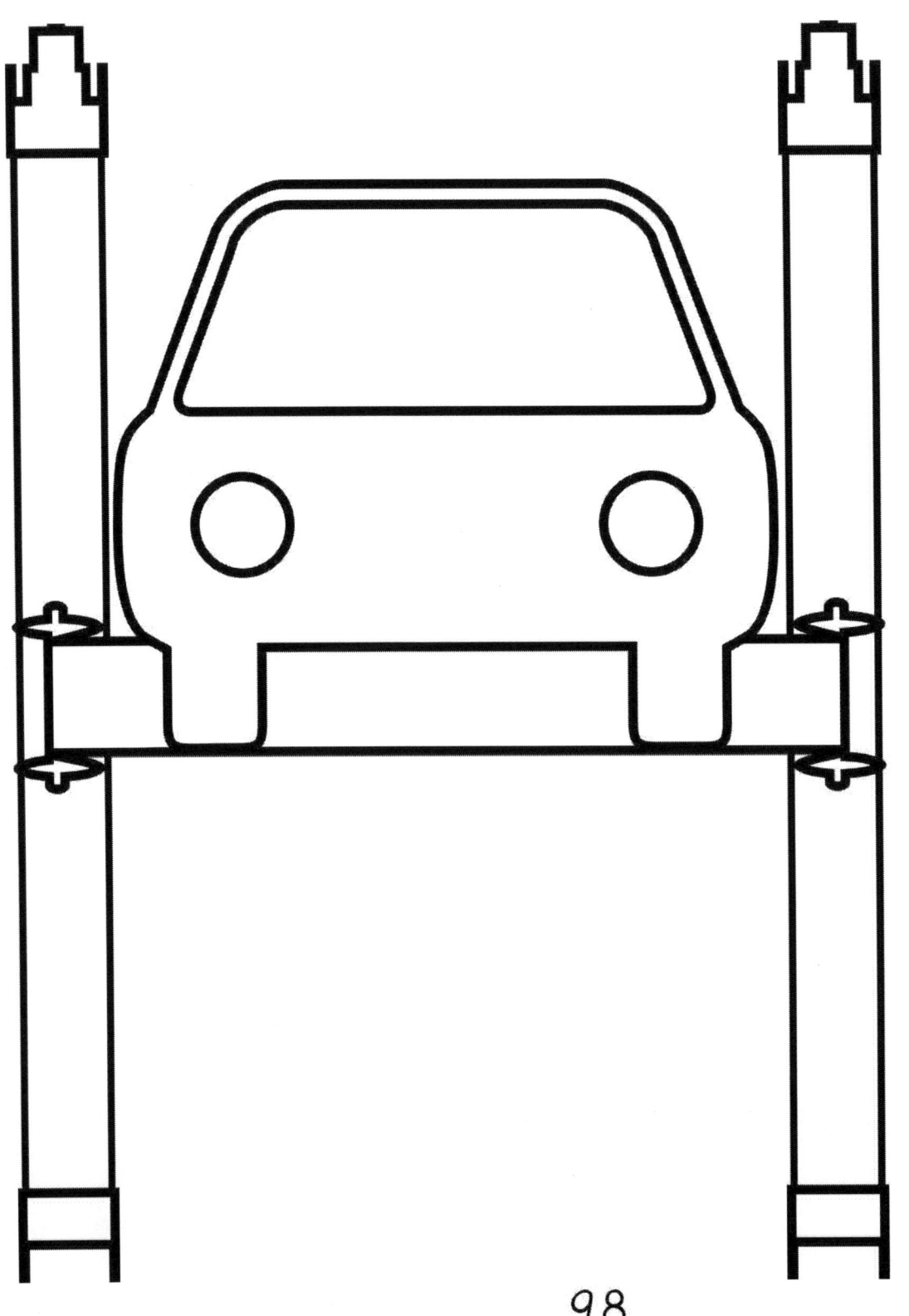

If the sedan is lowered to 0.5 meters off the ground, how much of that potential energy has been converted into kinetic energy?

BUT WAIT!

You might be thinking that energy and work have to do with more than just lifting and moving objects. And you're right! There are many types of energy, but our definitions and principles still apply.

Energy is defined as the capacity to do work, and work is force times the displacement through which it was exerted. How can this apply in other types of energy?

Consider water. When it is in the form of ice, the molecules are all very close together.

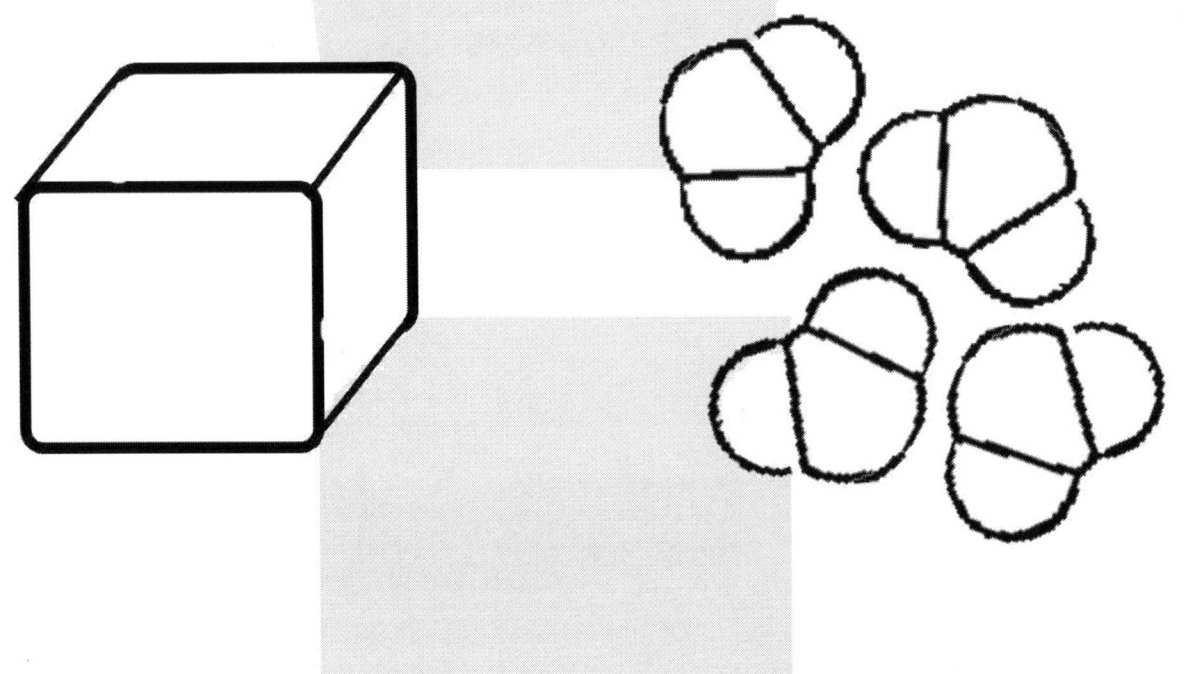

If you set an ice cube on your dining room table, over time it will melt. Heat energy is transferred to it from the table and the surrounding air.

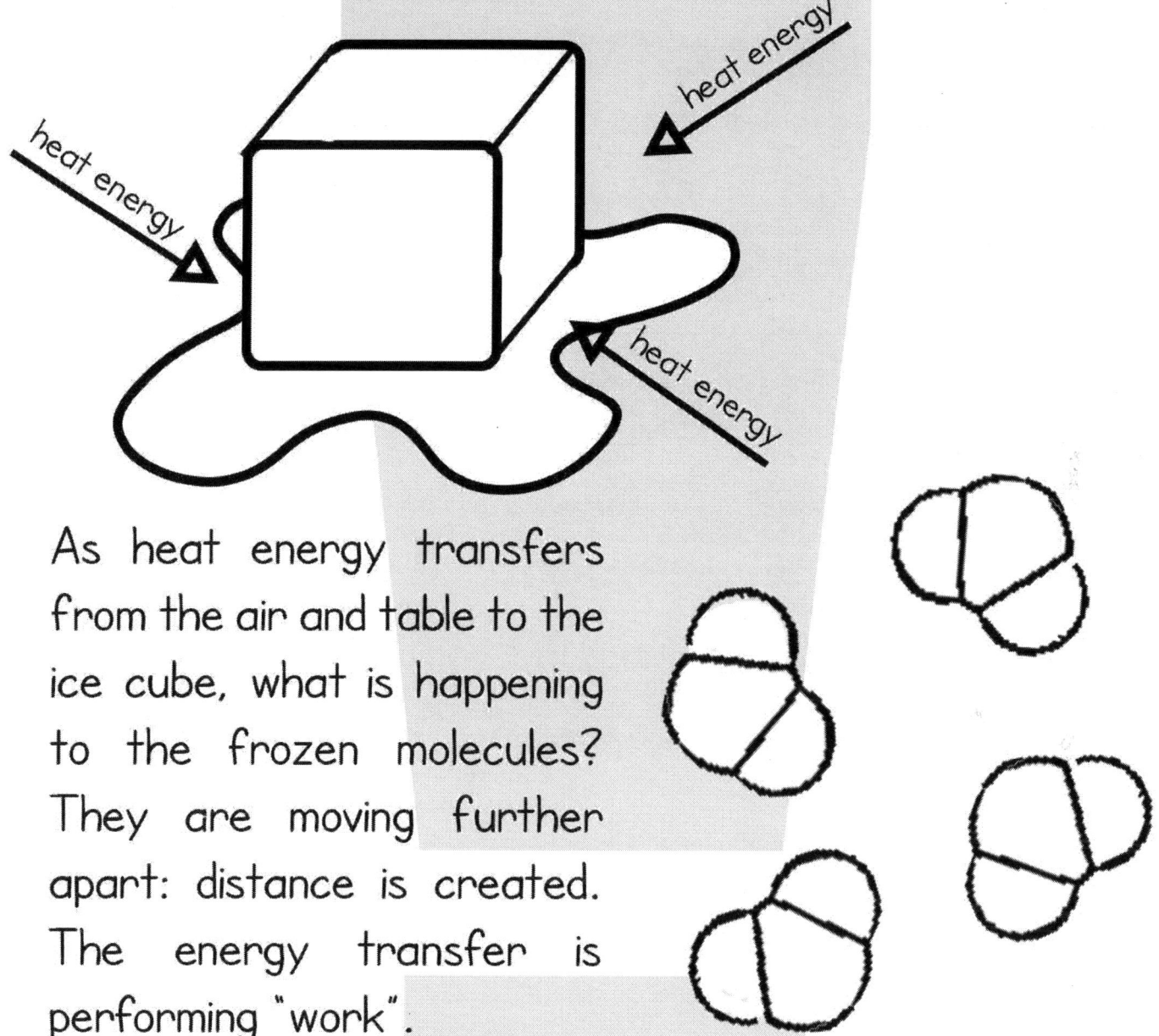

As heat energy transfers from the air and table to the ice cube, what is happening to the frozen molecules? They are moving further apart: distance is created. The energy transfer is performing "work".

As you can see, other forms of energy do exist, and they still fit the defintion of energy. Going into the other energy forms in depth, however, is beyond the scope of this book.

If work is force times distance, what is the difference between work done in a short period of time and work done over a longer period of time? A machine lifting 50lbs of bricks a distance of 2 meters in 40 seconds does the same amount of work as a machine taking 35 minutes to lift 50lbs of bricks the same distance. Does the amount of time in which work is done determine anything? Yes! The quantity of work done over time is called Power.

$$P = W/t$$
$$P = Fd/t$$

Power is measured in joules per second. One joule/second is called a watt.

$$1 \text{ joule/second} = 1 \text{ watt}$$

In the above example, did the first machine or the second machine use more power?

Using what you know about work and force, determine how much power it would take to move a 14,000kg vehicle across 80 kilometers in an hour and fifteen minutes.

Review!
Here are the formulas you've learned:

$$E = W = Fd$$

$$K.E. = \frac{1}{2}mv^2$$

$$P.E. = mgh$$

$$P = W/t$$

$$P = Fd/t$$

The dog wants to jump over the fence. The fence is 1.8m high and the dog weighs 42 newtons. How much energy will it take for the dog to jump the height of the fence?

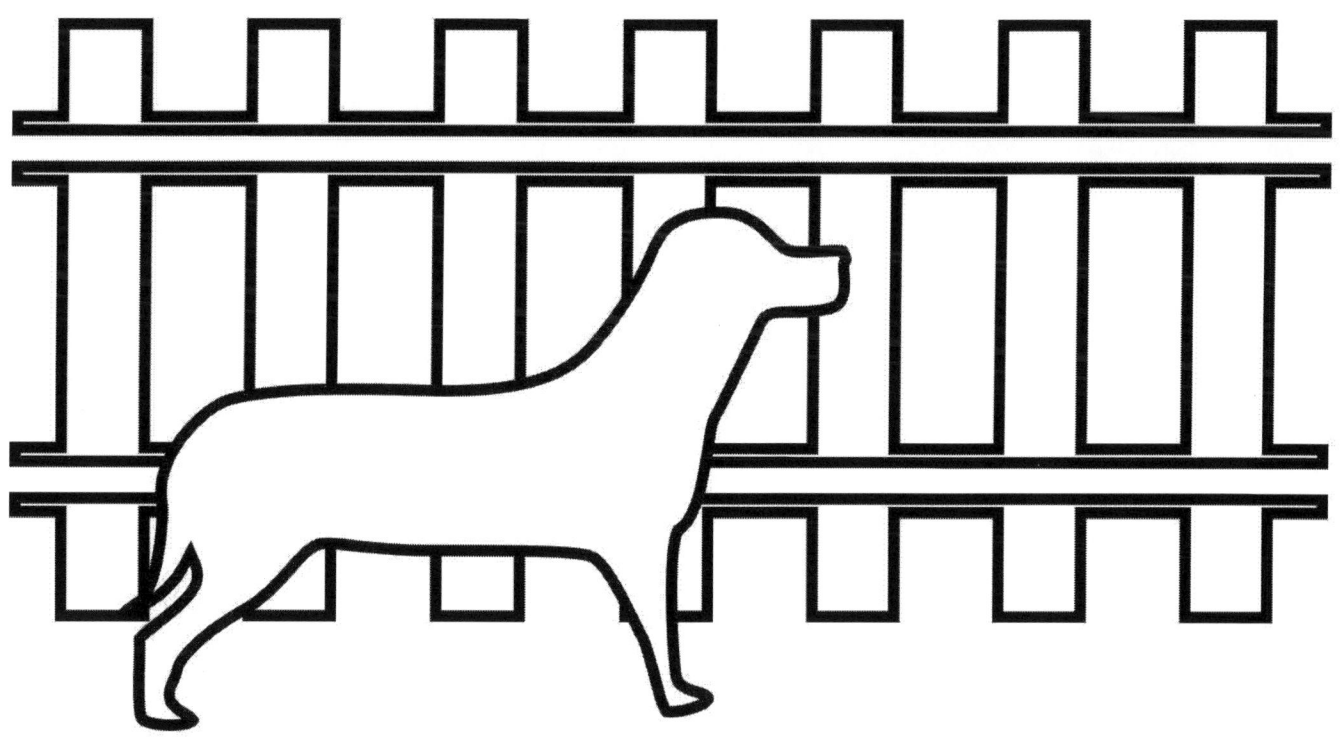

The table is 0.92 meters tall and the phone has a weight of 16 N. What is the phone's potential energy in relation to the floor?

It takes 23 newtons of force to lift the umbrella. How much work will it take to lift it 1.7 meters? Is the act of lifting the umbrella exerting kinetic energy or potential energy.

Page 1

1. 20m
2. 5m
3. draw runner 2 at the 25m line
4. draw runner 5 at the 10m line

Page 2

1. 2 minutes
2. 300 seconds
3. 90 minutes
4. 6 hours

Page 9

Runner 1: 10m
Runner 2: 9m
Runner 3: 12m
Runner 4: 7m
Runner 5: 13m

Page 10

Runner 1: 3.33m/s (slower)
Runner 2: 5.33m/s (faster)
Runner 3: 4.33m/s (slower)
Runner 4: 4.33m/s (faster)
Runner 5: 5.66m/s (slower)

Page 10

Runner 1: 3.33m/s (slower)
Runner 2: 5.33m/s (faster)
Runner 3: 4.33m/s (slower)
Runner 4: 4.33m/s (faster)
Runner 5: 5.66m/s (slower)

Page 11

Runner 1: 4.64s (fourth place)
Runner 2: 2.15s (third place)
Runner 3: 1.95s (second place)
Runner 4: 5.08s (fifth place)
Runner 5: 1.2s (first place)

Page 18

1. $(35m/s)/12s = 2.91m/s^2$
2. Change in velocity:
$0.48m/s^2(9s) = 4.32m/s$
Final velocity: 17.32m/s
3. $(42m/s-8m/s)/0.33m/s^2 = 103.03s$

Page 19

1. $(72\text{m/s}-10\text{m/s})/51\text{s} = 1.22\text{m/s}^2$
2. Change in velocity:
$0.15\text{m/s}^2(25\text{s}) = 3.75\text{m/s}$
Final velocity: 20.75m/s
3. $(80\text{m/s}-62\text{m/s})/0.61\text{m/s}^2 = 36.07\text{s}$

Page 25

1. $2(12\text{m})/5\text{s}^2 = 0.96\text{m/s}^2$
2. $\text{sqrt}(2(16\text{m})/1.34\text{m/s}^2) = 23.88\text{s}$
3. $1/2(4.6\text{m/s}^2)[(11.8\text{s})^2] = 320.25\text{m}$

Page 29

1. $(10.8\text{m/s})(24\text{s})+1/2(0.16\text{m/s}^2)[(24\text{s})^2] = 305.28\text{m}$
2. $[2(305.28\text{m}-10.8\text{m/s}(24\text{s})]/(24\text{s})^2 = 0.16\text{m/s}^2$
3. $[-10.8\text{m/s}+\text{sqrt}((10.8\text{m/s})^2 + 2(0.16\text{m/s}^2)(305.28\text{m}))]/0.16\text{m/s}^2 = 24\text{s}$

Page 31

1. circle: distance, average velocity
underline: time
$t=d/v=600\text{m}/(0.68\text{m/s})=882\text{s}$
2. circle: distance, starting velocity, acceleration
underline: time
$t=[-\text{vstart}+\text{sqrt}((\text{vstart})^2+2(ad))]/a$
$t=[-0.68\text{m/s}+\text{sqrt}((0.68\text{m/s})^2+2(0.07\text{m/s}^2)(550\text{m}))]/0.07\text{m/s}^2=116\text{s}$

Page 32

1. circle: time, starting velocity, acceleration
underline: change in velocity, final velocity
change in velocity = at = $(0.07m/s^2)(28s)$ = 1.96m/s
starting velocity + change in velocity = final velocity
0.68m/s+1.96m/s=2.64m/s
2. circle: time, starting velocity, final velocity, change in velocity, acceleration
underline: distance
d=vstart(t)+1/2at^2
0.68m/s(28s)+1/2$(0.07m/s^2)(28s)^2$=46.48m
OR
d=(vstart+(vchange/2))t
(0.68m/s+[(1.96m/s)/2])28s=46.48m
46.48m+50m=96.48m
600m-96.48m=503.52m

Page 33

1. circle: time, distance, starting velocity
underline: acceleration
a=2d/t2=2(503.52m)/38s^2=0.697m/s^2
2. circle: time, acceleration, starting velocity
underline: change in velocity, average velocity
change in velocity = at
0.697m/s^2(38s)=26.48m/s
average velocity = starting velocity + (change in velocity/2)
(2.64m/s)+(26.48m/s)/2=15.88m/s
3. circle: distance, time
underline: average velocity
v=d/t=72m/38s=1.89m/s
You have to run faster than 1.89m/s

Page 34

1. circle: starting velocity, acceleration, time
underline: distance
$d=1/2at^2$
$d=1/2(0.5m/s^2)[(3.8s)^2]=3.61m$
2. circle: starting velocity, distance, time
underline: acceleration
$a=2d/t^2=2(15m)/[(4.3s)^2]=1.62m/s^2$
3. circle: velocity, time
underline: distance
$d=vt=1.8m/s(2.6s)=4.68m$

Page 36

1. circle: starting velocity, final velocity, time
underline: acceleration
a=(final velocity-starting velocity)/t
$(4m/s-2m/s)/16s=0.125m/s^2$
2. circle: acceleration, distance
underline: time
$t=sqrt(2d/a)=sqrt(2(100m)/0.125m/s^2)=40s$
3. circle: starting velocity, final velocity, acceleration
underline: time
t=(final velocity-starting velocity)/a
$(3m/s-0m/s)/0.5m/s^2=6s$

Page 45

1. negative velocity

Page 58

1. $F=ma=(512kg)(1.2m/s^2)=614.4N$
2. $a=F/m=17N/36.4kg=0.47m/s^2$
3. $m=F/a=6.2N/0.82m/s^2=7.56kg$

Page 72
Earth's force pulling down on the box
The box's force pulling up on the Earth
$m=F/g=14.2N/9.8m/s^2=1.45kg$

Page 73
Dotted line in a straight line
Water pulled toward the moon on the side closest the moon, water further from Earth on the side furthest from the moon.

Page 74
Arrows pointing up:
The scale's force on the Earth
The cat's force on the scale
The cat's force on the Earth
Arrows pointing down:
The Earth's force on the scale
The Earth's force on the cat
The scale's force on the cat
Circle:
The Earth's force on the cat
$F=ma=22.1kg(9.8m/s^2)=216.58N$

Page 91
1. $W=Fd=15N(1.6m)=24$ joules
2. $W=Fd=30N(1.6m)=48$ joules

Page 92
1. $W=Fd=15N(12m)=180$ joules
2. $W=Fd=30N(6m)=180$ joules

Page 96

K.E.$=1/2mv^2$

$1/2(10,000kg)([(60km/hr)(1000m/km)(1hr/3600s)]^2)=138,888$ joules

Page 98/99

1. P.E.$=mgh$

$10,000kg(9.8m/s^2)(1.6m)=156,800$ joules

2. $mgh=10,000kg(9.8m/s^2)(0.5m)=49,000$ joules

Page 102

The first machine

Page 103

$F=mg=14,000kg(9.8m/s^2)=137,200N$

$t=75$ minutes$(60s/1minute)=4500s$

$d=80$ km$(1000m/km)=80,000m$

$P=Fd/t$

$(137,200N)(80,000m)/4500s = 2,439,111$ watts

Page 105

$E=Fd=42N(1.8m)=75.6$ joules

Page 106

P.E.$=mgh$

Weight$=mg$

P.E.$=16N(0.92m)=14.72$ joules

Page 107

$W=Fd=23N(1.7m)=39.1$ joules

Kinetic energy

Printed in Great Britain
by Amazon